down until it is snug in the Hirose connector, and tighten the included nuts so the Edison module doesn't fall off the breakout board.

Connect the Grove shield to the breakout board just as you would an Arduino shield and secure the connection between the two boards.

2. Intel Edison Software and Firmware Setup

Before writing the project source code, there are two software packages you need to download: the latest firmware for Intel Edison and a customized Arduino IDE for your computer.

Conveniently, there is a Windows installer for both here: software.intel.com/en-us/iot/hardware/edison/downloads. If you use OSX or Linux, simply download the Yocto complete image from the Board Firmware Software section on the same page. The second download you need to grab is the custom Arduino IDE for your OS, which can be found on the main IDE landing page: software.intel.com/iot/downloads.

3. Flashing the Intel Edison Board

Connect the Intel Edison board using two separate USB type A to micro cables, and also make sure that the SW1 switch is toggled toward the micro USB ports.

Follow the instructions in the Windows installer, then skip to step 5. For Mac and Linux users, on your computer, extract the firmware zip file that you downloaded and copy the contents of the folder to the Edison drive on your desktop.

4. Finishing the Flash

To complete the flashing process you'll need to log into the Edison using a terminal program and execute a command.

Grab a copy of CoolTerm and install it on your computer — freeware.the-meiers.org.

Set the baud rate for 115200 and the port as usbserial-* and be aware that the * will be a string of characters, but it is system dependent.

Initiate a connection with Edison by clicking the connect button. Wait a few seconds and hit return a few times. You should see the activity lights TX and RX flashing in CoolTerm.

After Linux fully boots, you will be

presented with a login prompt. Type in root and hit enter. By default there is no password set for the root user on an Edison.

Then type in the command: reboot ota . This command will overwrite all of the files and configuration scripts on your Edison, so make sure you've made backups of any specialized customizations you've made.

Once you press return, the system will reboot and load the new version of the firmware. You can verify the changes by cat /etc/version .

5. Grove Starter Kit

Grove sensors are an easy way to add data into an Intel Edison project. You'll be using temperature, light, moisture, and UV sensors in this build. All four of these are analog sensors and will connect to pins A0-A3. Go ahead and plug them in now.

6. Fire up the Arduino IDE to Code

Load the Intel version of the Arduino IDE that you download in step 2. You'll be writing all the code for the garden minder in this IDE.

Make sure the Intel Edison board is not mounted as a drive in your file system. Transferring sketches from the IDE to Edison simply will not work unless the drive is unmounted.

7. Writing the Sketch

Since there are four sensors that report unique values, there are four primary sections of code. As with all Arduino sketches, you will need to configure inputs and variables in your setup() function.

8. Monitoring the Plant

While the Intel Edison board is still connected to your computer and the Arduino IDE is running, select the Serial Monitor from the Tools menu and watch your plant's environmental conditions.

Now that you are comfortable using the four analog sensors and reporting values to the Serial Monitor, try to extend the features of the project. Try creating a graphical interface, or an auto-tweet function, or even adding a servo to let it water itself. The options are limitless — experiment and have fun. ●

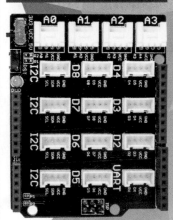

» **Intel Edison kit for Arduino** includes Edison
» **Grove starter kit**
» **Moisture sensor** Grove compatible
» **UV sensor** Grove compatible
» **USB A to USB Micro** (2)
» **Computer**

David Scheltema

CONTENTS

58

30

ON THE COVER:
Roll over, R2-D2 — BB-8 is the new droid everyone wants to build. Photo: Hep Svadja

18

42

Vol. 46, July 2015. Make: (ISSN 1556-2336) is published bimonthly by Maker Media, Inc. in the months of January, March, May, July, September, and November. Maker Media is located at 1160 Battery Street, Suite 125, San Francisco, CA 94111, 877-306-6253. SUBSCRIPTIONS: Send all subscription requests to Make:, P.O. Box 17046, North Hollywood, CA 91615-9588 or subscribe online at makezine.com/offer or via phone at (866) 289-8847 (U.S. and Canada); all other countries call (818) 487-2037. Subscriptions are available for $34.95 for 1 year (6 issues) in the United States; in Canada: $39.95 USD; all other countries: $49.95 USD. Periodicals Postage Paid at Sebastopol, CA, and at additional mailing offices. POSTMASTER: Send address changes to Make:, P.O. Box 17046, North Hollywood, CA 91615-9588. Canada Post Publications Mail Agreement Number 41129568. CANADA POSTMASTER: Send address changes to: Maker Media, PO Box 456, Niagara Falls, ON L2E 6V2

Make:®

> "If you don't know where you are going, you will wind up somewhere else." — Yogi Berra

EXECUTIVE CHAIRMAN
Dale Dougherty
dale@makermedia.com

CEO
Gregg Brockway
gregg@makermedia.com

CFO
Todd Sotkiewicz
todd@makermedia.com

CRO
Ed Delfs
ed@makermedia.com

CREATIVE DIRECTOR
Jason Babler
jbabler@makezine.com

EDITORIAL

EXECUTIVE EDITOR
Mike Senese
mike@makermedia.com

PRODUCTION MANAGER
Elise Tarkman

COMMUNITY EDITOR
Caleb Kraft
caleb@makermedia.com

PROJECTS EDITORS
Keith Hammond
khammond@makermedia.com
Donald Bell
donald@makermedia.com

TECHNICAL EDITORS
David Scheltema
Jordan Bunker

EDITOR
Nathan Hurst

EDITORIAL ASSISTANT
Craig Couden

COPY EDITOR
Laurie Barton

LAB MANAGER
Marty Marfin

EDITORIAL INTERNS
Sophia Smith
Nicole Smith

CONTRIBUTING EDITORS
Stuart Deutsch, William Gurstelle, Nick Normal, Charles Platt, Matt Stultz

CONTRIBUTING WRITERS
Taylor Alexander, John Baichtel, Doug Bradbury, Larry Cotton, DC Denison, Nick Dragotta, Jason Fabbri, Sam Freeman, Paul Gentile, Charles Guan, Parker Jardine, Bob Knetzger, Casey Kuhns, Nikos Mavrivakis, Jim McGreen, Forrest M. Mims III, Laura Rena Murray, Agnes Niewiadomski, Peter Oliver, Christian Poulsen, Benjamin Preston, Sean Ragan, Dan Rasmussen, Matt Richardson, Jason Poel Smith, Dave Southall, Chris Yohe, Kurt Zimmerman

Comments may be sent to:
editor@makezine.com

Visit us online:
makezine.com

Follow us on Twitter:
@make @makerfaire
@craft @makershed

On Google+:
google.com/+make

On Facebook:
makemagazine

DESIGN, PHOTOGRAPHY & VIDEO

ART DIRECTOR
Juliann Brown

DESIGNER
Jim Burke

PHOTOGRAPHER
Hep Svadja

VIDEO PRODUCER
Tyler Winegarner

VIDEOGRAPHER
Nat Wilson-Heckathorn

MAKEZINE.COM

DESIGN TEAM
Beate Fritsch
Josh Wright

WEB DEVELOPMENT TEAM
Clair Whitmer
Bill Olson
David Beauchamp
Rich Haynie
Matt Abernathy

CONTRIBUTING ARTISTS
Andrew Albosta, Matthew Billington, Bob Knetzger, Rob Nance, Zachary T. Nguyen, Andrew J. Nilsen, Damien Scogin, Scott D W Smith, Charles Platt

ONLINE CONTRIBUTORS
Cabe Atwell, Gareth Branwyn, Josh Burker, Tom Burtonwood, Lewis Calloway, Chandi Campbell, Jon Christian, Jeremy Cook, Jimmy DiResta, Josh Ellingson, Robert Foster, Matt Freund, Travis Good, Ronald Jaramillo, Erin Kennedy, Erica Lanotti, George LeVines, Goli Mohammadi, Jenn Nowicki, Haley Pierson-Cox, Marie Reedell, Phillip Rossoni, Andrew Salamone, Eric Schemplefeng, Leo Southam, Lori Stahl-Van Brackle, Theron Sturgess, Andrew Terranova, Erin Tomson, Michael Weinberg, Igor Vichikov

ENGINEERING INTERNS
William Felker, Adam Lukasik, Rachel Rosenson

VICE PRESIDENT
Sherry Huss
sherry@makermedia.com

SALES & ADVERTISING

SENIOR SALES MANAGER
Katie D. Kunde
katie@makermedia.com

SALES MANAGERS
Cecily Benzon
cbenzon@makermedia.com
Brigitte Kunde
brigitte@makermedia.com

STRATEGIC PARTNERSHIPS
Angela Ames

CLIENT SERVICES MANAGERS
Miranda Mota
Margaux Ryndak

MARKETING COORDINATOR
Karlee Vincent

COMMERCE

GENERAL MANAGER OF COMMERCE
Sonia Wong

RETAIL CHANNEL DIRECTOR
Kirk Matsuo

ASSOCIATE PRODUCER
Arianna Black

E-COMMERCE MANAGER
Michele Van Ruiten

MARKETING

VICE PRESIDENT OF CORPORATE MARKETING
Vickie Welch
vwelch@makermedia.com

DIRECTOR OF DIGITAL MARKETING
Patrick McCarthy

CUSTOMER RETENTION MANAGER
Heather Harmon Cochran
heatherh@makermedia.com

MARKETING PROGRAMS MANAGER
Suzanne Huston

DIGITAL MARKETING COMMUNICATIONS MANAGER
Brita Muller

MARKETING EVENTS MANAGER
Courtney Lentz

SOCIAL MEDIA MARKETING MANAGER
Jessie Wu

BOOKS

PUBLISHER
Brian Jepson

EDITORS
Patrick Di Justo
Anna Kaziunas France

MAKER FAIRE

PRODUCER
Louise Glasgow

PROGRAM DIRECTOR
Sabrina Merlo

MARKETING & PR
Bridgette Vanderlaan

CUSTOM PROGRAMS

DIRECTOR
Michelle Hlubinka

CUSTOMER SERVICE

CUSTOMER SERVICE REPRESENTATIVES
Kelly Thornton
Ryan Austin

Manage your account online, including change of address:
makezine.com/account
866-289-8847 toll-free in U.S. and Canada
818-487-2037,
5 a.m.–5 p.m., PST
cs@readerservices makezine.com

PUBLISHED BY

MAKER MEDIA, INC.
Dale Dougherty

Copyright © 2015
Maker Media, Inc.
All rights reserved.
Reproduction without permission is prohibited.
Printed in the USA by Schumann Printers, Inc.

CONTRIBUTORS

If you were going to build your own transportation solution, what would it be?

Taylor Alexander
Santa Clara, California [3D-Print a Badass R/C Race Car]

I'd build a bike highway system on platforms in Silicon Valley so I could ride a bike from Apple to Adobe and from eBay to Google.

Benjamin Preston
Brooklyn, New York [Own Your Car]

I'd use all the poor, neglected derelict vehicles in this country as a platform for some crazy new magnetic pulse propulsion technology.

Agnes Niewiadomski
Waterloo, Ontario [Floating Fantasy]

My invention would be a "peace pod," a rejuvenating cocoon that provides your body with everything it needs, so you arrive ready to seize the day.

Nikos Mavrivakis
Patras, Greece [Waterproof Bike Light]

I would make a bicycle frame from wood and rubbish, casting aluminum and plastic out of water bottles and soda cans for the connections.

Zachary T. Nguyen
Quincy, Massachusetts [Photos, World's Cutest Go-Kart]

I always dream of constructing a jetpack with two turbines to get unique angles for photographs.

PRINTED WITH SOY INK

Drawing Inspiration from
Maker Faire

Hep Svadja

GROWING UP WITH MAKER FAIRE

My son, Chris, was 8 at the first Maker Faire. He's now graduating high school and has been accepted to several universities (Duke, Ketting, University of Utah, and BYU). Field of study? Manufacturing and engineering. Why? Well, mostly because of *Make:* magazine (I have them all from the beginning), Maker Faire (we've been to 7 of the 10), and a bit of stuff in between that we got involved in because of the ideas and inspirations from the blog and the magazine. I've appreciated our partnership over the years.   –*Mike Johnson, Rocklin, California*

THE REAL COST OF A PROJECT

In issue 44 of *Make:* magazine, Mark Harrison tells us how to build a noodle copter! The article (page 42) says it cost $8 to buy all the materials, but my dad says it will cost over $100. Can you please tell me the actual cost of the copter so my brother and I can build it? Thank you very much.   –*Max Alexander, age 10*

EXECUTIVE EDITOR MIKE SENESE RESPONDS:

Max, sorry about any confusion with the Noodle Copter — the $8 price we posted is just for building this as the airframe. The electronics are separate, and yes, will cost about $100–$200 to get off the ground. I wish they didn't cost this much, but thankfully the prices are constantly dropping. We'll be more clear in the future.

SOCIAL MEDIA LOVE FOR MAKER FAIRE

➤➤ We were there Saturday and Sunday this year and were still wishing we'd had more time! So much to see, so little time! But, WE LOVED EVERY MINUTE OF IT! So much creativity in one place, AMAZING! Already looking forward to next year and the "Minis" along the way!!!
    –*Maryann Weigant, via Facebook*

➤➤ Thank YOU for inspiring a generation of doers, dreamers, and imagineers to make whatever they want to in this world. You have made my teaching career a whole lot more fascinating.
    –*Clint Johns, via Facebook*

➤➤ Thank you Maker Faire Team!!!! Your tireless commitment to the vision of *Make:* is an inspiration! And you all are a bunch of really great people too!!!! Thanks for having me, and thanks for all your hard work! What a great 3 days at the faire!!!!!!
    –*Scatha G Allison of Miss Velvet Cream Neo Couture Clothing, via Facebook*

➤➤ Thanks for putting this event together. Participation and variety were much better than 2014. Looking forward to next one already. Oh BTW my kids were so tired at the end — they are asking for more days of Maker Faire so they can keep coming back to check out other zones. I agree.
    –*Baris Eris, via Facebook*

➤➤ We drove down all the way from Eugene, Oregon for this. Second-year attendees. I want to expose my daughter, age 4, to this awesomeness so she too can be inspired. Thank you Makers!
    –*Jen Hi, via Facebook*

➤➤ What an awesome, inspiring, rejuvenating, exhausting, fun, invigorating weekend! Met so many fabulous people that we're planning on contacting again very soon. Thanks so much!
    –*Carol Maley of Marin Maker Mobile, via Facebook*

Make: Amends

In Volume 45's "Who's Watching You" (page 70), the LibraryBox File Sharer should have been credited to Jason Griffey, the creator and author of the project. The accompanying photo on page 71 should also be credited to Griffey. Our apologies, Jason!

Re-Inventing the First
Text Message System

See the relay, Page motor, and experiences
from students at makezine.com/46/welcome

BY DALE DOUGHERTY, founder and Executive Chairman of Maker Media.

ON A VISIT TO SUTHERLAND MIDDLE SCHOOL IN ALBEMARLE
COUNTY, VIRGINIA, I SAW GROUPS OF STUDENTS BUILDING
a motor out of 3D-printed and laser-cut parts plus coat-hanger
wire. Up on the screen at the front of the class was a 3D simulation
of the motor. The teacher, Robbie Munsey, explained that the
students were re-creating a Page motor, named after its inventor,
Charles Graton Page, and that the invention was patented in 1854.
"Looking at historical inventions is a way for students to interact
with technology and understand it," says Munsey, who, dissatisfied
with the science curriculum, wanted to employ more hands-on
learning. A Page motor makes it clear what a motor does and how
it works, and actually building it brings this home to students.

Initially, Munsey struggled with hard to obtain supplies for more
complex projects. He met Glen Bull, a professor at the Curry
School of Education at the University of Virginia, who was talking
about 3D printing (see *Make:* Volume 41, "The Lab in the Classroom"). "He let me borrow a really old 3D printer," says Munsey.
"Glen said take it home and see what you can do it with it." He did,
and discovered that the printer was the answer to his supply problem. "You could build anything with it, and I noticed the kids were
mesmerized by it," he says.

Bull invited Munsey to join his online graduate class on edtech.
The next challenge in that class was to build a telegraph, and it
caught Munsey's interest. On his own, he built the telegraph. Bull
was ecstatic. He asked Munsey if he thought 8th grade students
could build one. "Absolutely," Munsey replied, and he recruited
an engineering teacher, Eric Bredder, who ran Sutherland's
makerspace, as a collaborator to make it happen.

Munsey believed that building a telegraph with 3D-printed parts
would be practical and meaningful to the students. "I told them
that the telegraph was the first text message system ever," he says.
"Modern-day relays are plastic boxes that you can't open up. Even
if you could open them, you wouldn't understand them. The great
thing about these inventions is that we can see how they work."

Munsey and Bredder gave the students original documents, such
as the patent application. Rather than starting with a kit or prepared
instructions, students had to learn how it worked and design their
own version. They built the telegraph with Autodesk 123D and a new
MakerBot. "All of us were surprised by our success," says Munsey.

Bull was so excited by the work that he suggested they show
some people at the Smithsonian in Washington, D.C. The class
created a presentation, and "blew them out of the water," says
Munsey. It resulted in a formal collaboration between the University
of Virginia, the Smithsonian, and Sutherland Middle School, and led
to an NSF grant.

Later, Bull reached out to Princeton's Michael Littman, an expert
on historical inventions, and he joined the collaboration, with a grad
student adding 3D model simulation.

The next challenge was the Page motor. "Tough," Munsey
describes it. "Crazy tough." But the students succeeded again.
"Having built the Page motor, students could learn what's great
about it and what's not so great about it," he says. "Then I asked
them, what can you do to change it? What would you do to make
it better? Can it look even more like the original Page motor?"
Munsey was delighted to hear the technical language seep into their
conversations — "the commutators have too much friction," "we're
pulling 3 amps, how do we lower that?" — it was no longer a science
project; it became their own project.

"I never once lectured them on any of this. Not a single direct
lesson. It wasn't me pushing curriculum to them; it was them
pulling the knowledge in," Munsey says. "The power of choice is
so incredible."

"Our kids said that by working with the original inventions and
studying what the inventors wrote, they honestly felt connected
to inventors like Morse," Munsey says. "When I pointed out that
these famous inventors didn't fully understand what electricity
was, I realized that neither do my students. The innovators and the
students have holes in their understanding, and we could talk about
that." They learned about the invention, but also how the minds of
inventors work a lot like their own. ⚙

Hep Svadja

The shape of things to come.

Visit World Maker Faire
New York on Sept. 26-27, 2015
to see the all-new UP BOX.
Learn more at UP3D.com

Sign up today!

Maker Camp
Make:

A free camp for building, tinkering, and exploring.
Online and in your neighborhood.

July 6 – August 14
makercamp.com

MADE ON EARTH

The world of backyard technology

Know a project that would be perfect for Made on Earth? Email us: *editor@makezine.com*

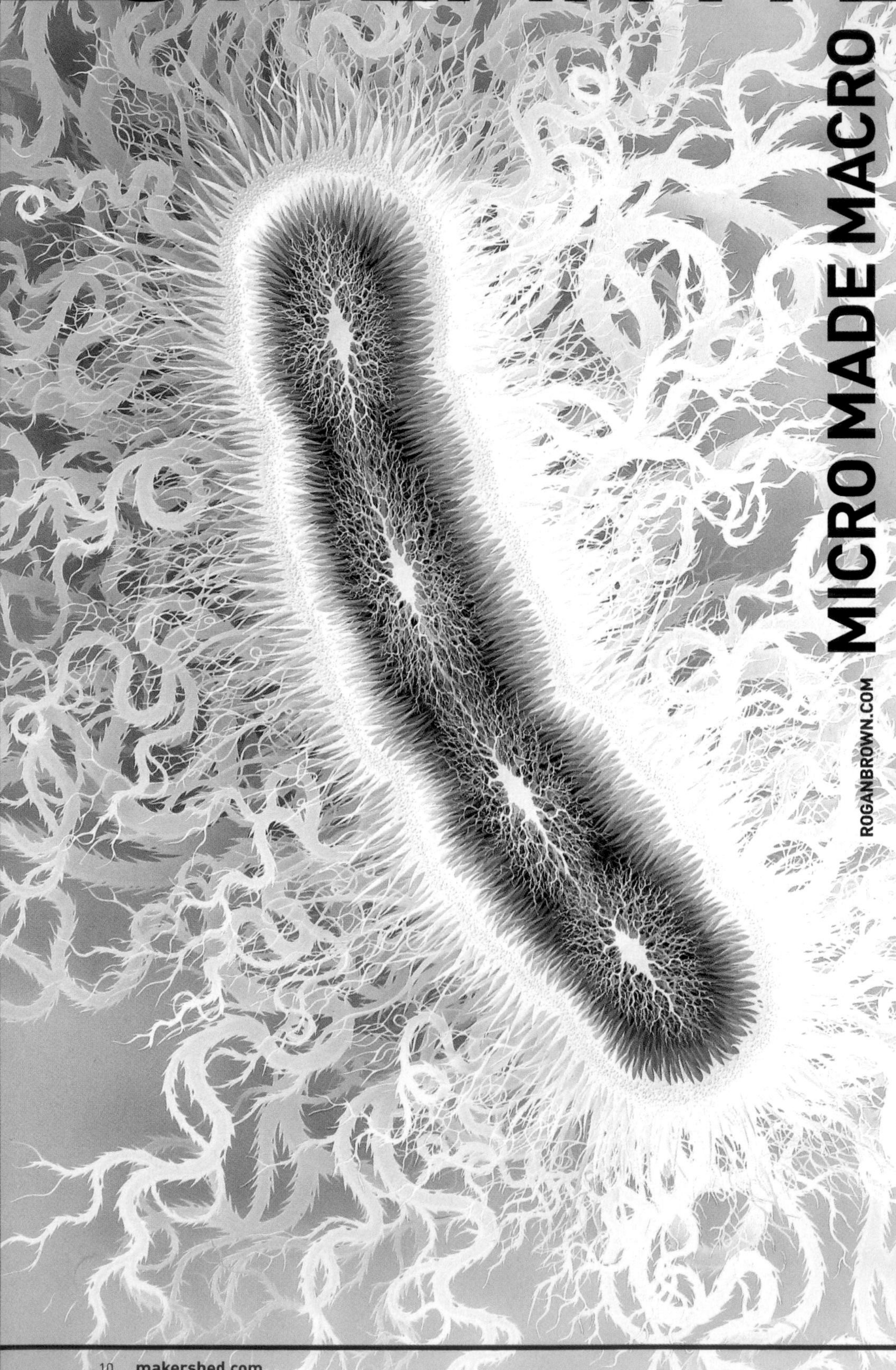

MICRO MADE MACRO

Rogan Brown makes four-dimensional paper sculptures of microbes. He calls himself a scientific surrealist, and the fourth dimension is time.

The time element the 48-year-old artist is referring to is the four to five months it takes to make the sculptures using his Epilog laser cutter. Brown prefers to use paper as his primary medium because "it's a humble material, everyone can access it … I use a populist technique that allows the public to access something that might be beyond their comprehension."

After living in London for many years, Brown relocated to a remote area bordering a national park in southern France. "I was trying to find a way into seeing what was surrounding me because the traditional art forms of representing nature didn't appeal," he says.

He purchased a microscope and found himself captivated by the precise observations he recorded in the environment around him. "Detailed scientific drawings of nature look completely surreal," Brown explains. "We live in a scientific age and artists engage in the dominant narrative."

Brown's microbe exhibit, *Invisible You*, will be on display for the next five years at the Eden Project, an educational charity in the U.K. He will also be exhibiting *Outbreak*, a huge installation of 800 microbes flowing out of a petri dome and crawling over the gallery walls, at The Coda Museum in the Netherlands this June through October.

— *Laura Rena Murray*

ROGANBROWN.COM

FLOATING FANTASY
BALLOONMANOR.COM

Drew Ripley is a full-time balloon twister. From birthday parties to major art projects, he lives to entertain crowds through the fine art of folding air. When Airigami's **Larry Moss** and **Kelly Cheatle** asked him to help on Balloon Manor, a large-scale ephemeral balloon installation, he teamed up with a crew of 60 to bring life back into an old department store in Rochester, New York.

The Amazing Air-Filled Under-Sea Adventure used more than 40,000 balloons to fill the building's five-story atrium. For four days, it was a frenzy of activity, as artists worked from blueprints, air compressors ran nonstop, and rigging was installed overhead to support it all.

Inflating that many balloons requires machinery to quickly pump the correct amount of air. Dissatisfied with the reliability and precision of commercial balloon pumps, Ripley and Moss built the Inflatinator, a programmable machine that was the only automated equipment used in the 2015 Balloon Manor.

They made their first prototype with a breadboard, an Arduino, and some solenoids packed into a plastic container, and then redesigned it into a safe and robust machine with the help of local engineer **Paul Walker**. The team continues to tinker with the functionality to increase productivity for future balloon builds.

"After working 16 hours straight I go home happy, exhausted, but still thrilled," says Ripley. "Everywhere I go, I get to give people that moment of whimsy and being free and recognizing that life is awesome, and what could be better than that?"

—*Agnes Niewiadomski*

Airigami

MAKER FAIRE BAY AREA 2015

MAKERFAIRE.COM

1

Hep Svadja

Evan Jones

2

3

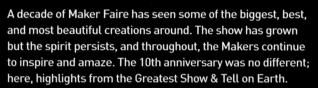

4

5

A decade of Maker Faire has seen some of the biggest, best, and most beautiful creations around. The show has grown but the spirit persists, and throughout, the Makers continue to inspire and amaze. The 10th anniversary was no different; here, highlights from the Greatest Show & Tell on Earth.

1. Robot Resurrection *towered over the South Lot*

2. Zolie on her Trilobite *car*

3. A glowing fashion show featured 3D-printed prosthetics

4. Praying with Fire *in front of* Rhino Redemption

5. Mission Delirium jams in front of Celestial Mechanica

6. James Peterson's Sessilanoid *responds to touch*

7. Paul Cesewski's Bicycle Ferris Wheel *is powered by riders*

8. A mini Tesla coil electrified the darkened Fiesta Hall

9. An army of R2-D2s prowled the grounds

6

7

8 9

Documenting Projects

with Sprout by HP

Written By *Make:* Editors

Make Your Own
Custom Stamp

GREAT PROJECTS NEVER APPEAR FULLY FORMED IN YOUR MIND; THEY TAKE WORK.
Developing the idea, spending hours building it, and wrapping it all up with a nice set of photos in a blog post isn't easy. For most people, going from idea to reality is enough to tackle, but for a truly great project, each step along the path should be documented.

Detailed step shots and explanations make it easier for people to replicate your success, and even improve on it. The best-documented projects are those that transcend a singular build and begin a life of their own. Alas, if documentation were that easy, then everyone would do it. We've all got our excuses ...

« "I don't want to stop what I'm doing to take pictures."
« "I don't have the right equipment ..."
« "I always forget to!"

Documenting a project adds a whole additional layer to an already complex process, and as a result, most people don't bother to do it. It's sad to think of all the fantastic builds that we've never seen, simply because people didn't pick up a camera and snap pictures along the way.

So how can we turn documentation into an activity that is practical for everyone, and that they might even enjoy? The best approach is to make it as seamless and easy as possible. If we all had machines that made it simple to plan, photograph, and share our build process, we'd all be much more likely to show the world what we've made. Thankfully, the Sprout by HP is the perfect tool for the job, and we're here to show you how to use it.

DEVELOPING YOUR IDEAS

Most projects start as a scribble on a napkin, a sketch in a notebook, or a collection of post-it note doodles. It's important to capture your ideas with whatever is nearby when inspiration strikes, but it can be tricky to keep track of the results. With the Sprout though, keeping all of your ideas in one place is a snap, no matter where they happen.

In the Create app, you can build a virtual project

Hep Svajda

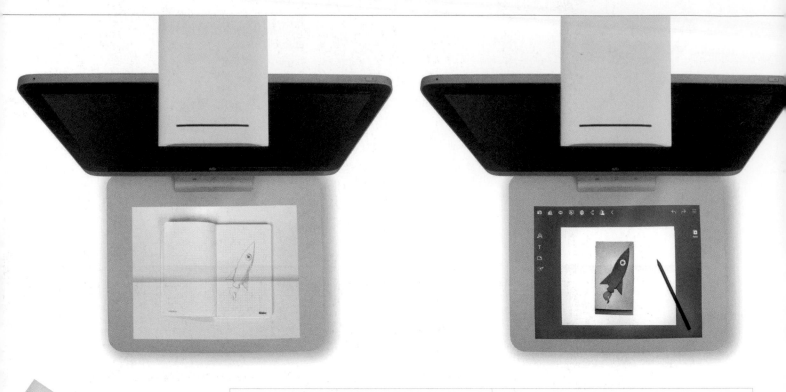

board to collect notebook sketches, inspirational images from the web, and draw new designs on the fly. The Sprout allows both digital and physical plans to live together in the same place. Having a single place to track all of your plans helps you organize your ideas.

CAPTURING THE PHYSICAL AS DIGITAL

Opening the Create app will bring you to a blank workspace. To scan in the contents of your notebooks, tap the camera icon in the upper left corner to open the Capture feature, and then place your notebook on the Touch Mat. Tap the large camera icon on the touchscreen, and a photo will be taken. If you're satisfied with the resulting picture, click the check mark. The image will be placed on it's own layer in the workspace, as well as in the image gallery for later use.

Of course, you're not limited to capturing just sketches or notes. You can also place parts and materials in your project board to help develop ideas further. If it can fit on the mat, the Sprout can snap a picture of it and place it in the workspace.

IMAGES FROM THE WEB

If inspiration strikes while browsing the web, you can simply copy and paste images directly into the workspace. Any image in the workspace can also be selected and rotated or resized with the standard multi-touch pinch, zoom, and rotate gestures you're used to. Tapping on the Edit tools in the menu on the right, you can even crop, remove the background, or punch out a shape in an image. These built-in tools make it easy to quickly modify the image just the way you want it.

DRAWING YOUR IDEAS

The Sprout comes with a stylus pen that makes it easy to sketch or take notes directly on the Touch Mat. Tapping the paintbrush icon will show the Ink tools. Here you can choose your pen style, the color you'd like to draw in, and the line width. The pencil will create lines of uniform thickness, while the calligraphy pen will produce a line of varying thickness based on the angle that you draw. And just as in the physical world, the highlighter will create areas of partial transparency for emphasis.

BUILDING

Once you begin building, the real work begins. Thanks to the durability of the Touch Mat, you can work on most small projects without worry. The Touch Mat's ultra-resistant coating deflects spills, cuts, glue, and ink, so it will stand up to most mishaps.

Working directly on the Touch Mat lets you take advantage of the Sprout's Capture application for mid-build photographs. When you start your project, open the Capture app and you'll be greeted with a large view from the camera above

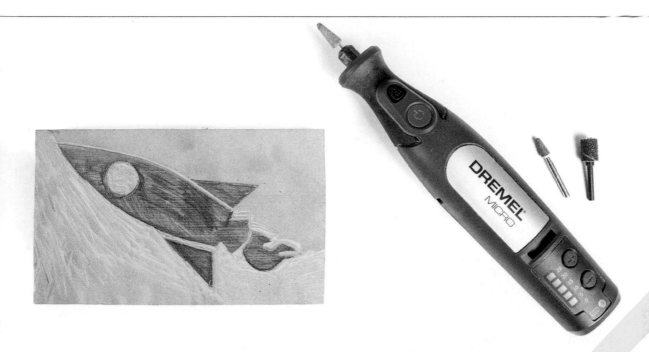

the Touch Mat. For the projects that are too large for the Touch Mat, you can switch to the camera in the touchscreen bezel by tapping the arrow icon in the bottom right corner of the screen.

TAKING PICTURES

To snap a photo of your project-in-progress, simply tap the camera icon on the touchscreen. When a photo is taken, by default the Sprout will separate all of the objects in the picture. This is great to capture all of the tools and materials used at the beginning of a build, but selecting the "separate objects" icon on the left side of the screen can disable the separation feature. Tapping the + symbol will save the picture and allow you to take another picture, while tapping the check mark will save the current picture and exit the app. All saved images are conveniently placed in the Gallery for later use.

CUT THE LIGHTS

Keeping the Capture app open will keep the Touch Mat display on, which might be distracting while you're working. Thankfully, turning the projection off is easy. Simply press and hold the touch-sensitive Touch Mat symbol on the silver base of the Touch Mat, and the Touch Mat display will be disabled. To turn it back on, press the same symbol, and the display will come back on right where you left off in the Capture app, ready to take another picture. As an added bonus, if you'd like some soft lighting while working, you can tap the front of the Illuminator arm to turn on the built-in overhead lamp.

SHOWING YOUR STEPS

Having shots of a project in progress can be useful, but to help people understand the process, adding a description of what was done in the picture is even better. When you're creating a set of steps for others to follow when replicating your build, the Sprout's Create app has all the tools you'll need to help you show your work.

With the Create app open, you can open the image gallery on the touchscreen and swipe through the pictures you took earlier during the build. To add an image from the gallery to your workspace, just swipe it down toward the Touch Mat, and it will appear within the workspace of the Create app. After resizing the image, you can use the Ink tools to draw and annotate directly on top of your image, or use the Text tools to add a description of what you were doing in the photo. When you're done, simply use the drop-down menu in the top right corner to save the file as a JPEG, PDF, or PNG, and open a new document to start designing the next page in your build steps.

For many people, project documentation is a pain, but it doesn't have to be. The Sprout makes it easy — even fun — to take great photos during your build, and create beautiful instructions for sharing your project with the world. ●

SPACE CHASE

THE FOUNDERS OF **CARBON ORIGINS** MOVED TO THE DESERT TO LAUNCH ROCKETS. IT WAS HARDER THAN THEY EXPECTED.

Written by Nathan Hurst

Kailey Shara, Danna Torio, Jaimie Hadden, and Amogha Srirangarajan pose with Carbon Origins' Phoenix 0.3 test rocket on the morning of its first — and last — launch.

IF you follow Aerospace Highway north out of Mojave, California, take a small road off to the east, continue as it becomes gravel, and then sand, and pass a dry lake bed, you'll end up at the Friends of Amateur Rocketry (FAR) launch site. These 10 acres of bare desert nestled between mountains and bordering a California desert tortoise preserve are scarred with blast marks from hundreds of experimental rocket launches.

It's here, on a hot, sunny Sunday in April, that Carbon Origins, a startup consisting of four recent Case Western Reserve University engineering students, has set up Phoenix 0.3, a 10-foot-long, 4-inch-diameter silver and black needle of a rocket. Inside are six commercial flight controllers, plus three of their unique, custom-built boards designed to track the rocket's flight past Mach 2.5 and up to 43,000 feet.

Disassembled, the rocket fits in the back of Danna Torio's Toyota Highlander, which is how Carbon Origins got the rocket out to the FAR site from their house/corporate headquarters in nearby California City. Amogha Srirangarajan stands beside the rocket on the platform, arming the electronics manually. They beep through the fiberglass body in acknowledgment.

"She'll be dented and scratched when she comes back," he says, affectionately.

A few feet away, Kailey Shara makes sure the GPS unit is locked onto satellites. Then everybody retreats to a nearby bunker, and Srirangarajan video-calls Peter Dixon, their fourth co-founder, who could not be present.

Nobody slept last night, Srirangarajan says, but Torio and Jaimie Hadden, who volunteered at the launch, admit to catching naps in the car.

The countdown begins, from 10. It reaches zero, and nothing happens.

A HOUSE IN THE DESERT

Carbon Origins can trace its trajectory to the rocketry club at Case Western in Cleveland. Srirangarajan founded the group, which expanded into one of the most popular clubs at the university.

The club went to Utah to launch an 18-foot, two-stage rocket. It blew up. More than $23,000 worth of school, sponsor, and personal money went up in smoke, and they didn't know why because the flight controller sensor array the rocket needed to tell them what happened didn't yet exist.

Eventually, they pieced it together, based on footage of the launch and an exhaustive process of picking up the scattered debris. The first stage lifted off fine. At best guess — an educated guess, but still not completely certain — the flight controller, which ignites the second stage, did so before stage 1 was finished. It wrecked the body, says Shara. The engine blew right threw it.

"It was a pretty ambitious 2-stage rocket," she says. "It was kind of pushing the capabilities of that flight controller."

Some in the club thought they could do better. Srirangarajan, Shara, Dixon, and Torio split off to start Carbon Origins, with the dual goals of making it easier for citizens to explore space and building the tools, including that missing flight controller, that would help them do that. The flight controller became a custom, durable, Arduino-compatible device called Apollo.

"That was the point at which we said, this is more than a hobby," says Srirangarajan, who is president and CEO of Carbon Origins. "From the ashes was born another rocket, and a company."

"That is our single goal, space, and we'll do everything we can to make sure that people like us, when they get to universities, and when they get to a part in their life where they have the basic skills to do something cool, they have this resource to put stuff in space," he says.

In the summer of 2014, they moved to California City, rented a four-bedroom house with a rocky yard and heavily taxed air conditioner, and set up shop. Literally — they built a shop in the living room, with a drill press, shop vac, MakerBot, a big monitor for CAD, bright work lights on stands, and a big bank of drawers containing bolts and other small parts. There are five whiteboards covered in calculations and diagrams, plus a fridge with three faces covered in dry-erase markings.

Five more whiteboards decorate the walls

"FROM THE ASHES WAS BORN ANOTHER ROCKET, AND A COMPANY."

Hep Svadja

Carbon Origins' multi-sensored Apollo board.

The recovered payload (above) and Phoenix 0.3 parts (below).

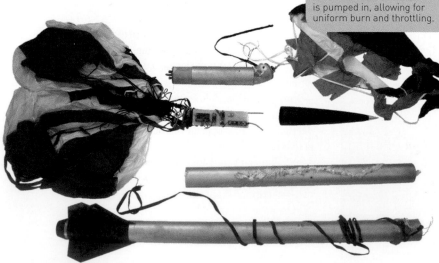

What's in a rocket engine?

There's a lot of effort going into making better rocket engines. Carbon Origins figured their resources were better spent on optimizing the rockets, and outsourcing the engine. But other teams of students and professionals are experimenting with different fuels. Here are the three most common types:

Solid Propellant: Black powder compressed with an oxidizing agent to provide oxygen, solid engines are designed not to explode but to burn uniformly and expel gases out the back. Some include a charge on burnout to eject a parachute, while others are built to ignite a second stage.

Liquid Propellant: Liquid-propelled engines mix a fuel like petroleum with an oxidizer in a combustion chamber. They're more complex than solid engines, but can be re-used, as well as throttled.

Hybrid: Typically, the fuel is solid while the oxidizer is pumped in, allowing for uniform burn and throttling.

of the garage, which has been converted to a white-walled office with fluorescent lights. There, the crew has a Form1 printer, vinyl cutter, DJI Phantom, cases of Red Bull, and inward-facing desks, like the bridge of a star cruiser. There's no garage door opener; the door itself has been insulated to resist the desert heat and sealed around the edges, though sand and bugs tend to slip in.

But maybe the most important feature of the house stands next to the kitchen. It is a 1:8 scale model of a payload-carrying, space-bound, two-stage rocket that is their ultimate goal. They don't know what'll go in it, but its purpose is to make it easier for Makers to launch stuff into space. "We built this so we could have a physical reminder of what we're aiming for," says Torio.

Nearly everything you'd need to build a rocket resides in the house with them. "Having an apartment like this, where you can wake up and walk straight to your shop, that's incredible," says Srirangarajan. They hardly even need to leave; Torio's car is the only one they have, and Srirangarajan says he seldom carries his wallet — it's rare that an expenditure isn't related to the business.

The lack of distraction is a big part of why they moved to California City, population 14,120. They don't exactly have a lot to do there, socially speaking. They go flying in a Cessna 172 (Dixon has been flying since he was 16). They take trips to Home Depot, race go-karts, pilot quadcopters, or just drive into the desert in different directions to see what's there. They order from Jesse's, one of two pizza joints in town. And the people they do socialize with tend to be space enthusiasts as well.

"Mojave is a whole other world," says Srirangarajan. "Every other person you meet is into space. There is an actual spaceport where we can go launch rockets into space if we wanted to, tomorrow."

"ROCKETS ARE HARD"

Model rocketry has been an educational staple for decades. Take a cardboard tube, stick on some fins, a nose cone, and an engine, and you've got a rocket.

"Model rockets use the same basic materials that have always been used," says David Raimondi, president of LUNAR, the Livermore (California) Unit of the

Amogha Srirangarajan

Duchesne Torio

National Association of Rocketry. "The two big changes are the complexity and size of the electronics available today, and the availability of motors."

For at-home model rocketry, many of those motors come from Estes Rockets. They're the little light-brown cylinders, around the size of a roll of pennies, filled with solid, pressed black powder rocket fuel — also known as gunpowder.

"Business is good. It's still very popular, but it has changed," says Mike Fritz, Estes' director of product development. "Today's consumers seem to be a little more interested in, I got an hour, what can you give me?"

Estes sells rocket engines up to size G. The engine inside Carbon Origins' Phoenix 0.3 is size O, sourced from Cesaroni Technology, an industrial manufacturing company in Canada. Since each letter increment represents a doubling of rocket power, the Phoenix 0.3 is more than 8,000 times more powerful than the one you likely launched in middle school.

Any motor larger than a G requires certification from the National Association of Rocketry or the Tripoli Rocket Association, though Fritz points out that, at that level, you're probably firing rockets beyond what you can see. That's one of the reasons onboard electronics can be useful. A radio or GPS for tracking means once it goes out of your sight, it's not gone for good. And an altimeter (usually barometric) can tell you exactly how high it went, something else that quickly gets pretty hard to distinguish by eye.

An O motor provides up to 40,960 newton-seconds of thrust. The O in Carbon Origins' Phoenix 0.3 is just over the cutoff from N size, with 21,062 newton-seconds — a lot for a 63.4-pound rocket. Enough, according to their calculations, to blast it to that 43,000 foot, Mach 2.5 goal.

The engine doesn't push the rocket the whole way; at around 6 seconds into the flight, says Srirangarajan, it should run out of fuel. But it'll still be traveling more than twice the speed of sound, and that momentum will carry it a great deal further, for another minute or more. As it reaches its peak, the onboard controller will recognize that air pressure has stopped changing, and deploy the CO_2 charge that will pop the

rocket open and eject the drogue parachute. This smaller chute will allow the rocket to descend quickly to an altitude of a few thousand feet, where a main chute will deploy to soften the landing. By the time it touches down, it should be going a nice, easy, 10mph.

That wasn't quite how it worked out for predecessor Phoenix 0.2. Launched in June 2014, its chutes deployed while it was still accelerating. The nylon strap tore through the aluminum body, and the rocket came down ballistic — in free fall. They recovered the warped, crumpled body, which stands near their kitchen beside the scale model. The flight controller showed that the air pressure remained constant, suggesting that the small hole in the rocket that allows pressure to equalize as it flies might have gotten plugged somehow. When the altimeter can't read the pressure change, it doesn't know when the rocket peaks, and therefore when to release the chute.

"Rockets are hard," says Shara, noting that in Phoenix 0.3, they've included redundant flight controllers and parachute ejection charges.

Phoenix 0.2's crash was just one of many things that can go wrong with rockets. The FAR site is run by a group of mentors from the rocketry industry, who allow university clubs to use it, and companies to launch for a small fee. Kevin Baxter, FAR's president, helped buy the land after he saw a need for a place students could go to launch rockets. The site lies under the Edwards Air Force Base umbrella, so there are no commercial flights going over. In 2003, FAR was founded as a nonprofit.

Students from all over southern California, the Southwest, and even as far as Annapolis, Maryland, launch twice a month at the site. Failures, sometimes including spectacular explosions, happen frequently. "Failed launches are simply learning experiences and they are a common occurrence at FAR," says Baxter. "Hence the viewing trenches and blockhouse."

On Saturday, about 24 hours before Carbon Origins was scheduled to launch, a rocket crashed. It was silver, red, and white, liquid oxygen-fueled, with a 3D-printed engine and a wide set of fins. It lifted off slowly, almost hovering in midair. It tipped sideways into the wind, flew a few

Interested in fueling your inner rocket scientist?

Check out our book *Make: Rockets*, from the Maker Shed (makershed. com). Learn the aerodynamics that influence how they fly, and try DIY projects from simple water rockets to two-stage payload carriers.

Hep Svadja

hundred feet, and hit the ground with a thud and a plume of dust.

Other rocket engineers were quick to point out what may have gone wrong. The launch rail was short, one said, especially in relation to the large fins. With a short rail, the rocket doesn't achieve as much speed before it's free, and with large fins, it's liable to tilt into the wind, like a weather vane.

It's a problem with stability, which is primarily dependent on the distance between the center of mass (determined by weight) and the center of pressure (determined by airflow over the shape of the rocket, especially the fins). The farther the two are apart, the more stable the rocket. The large fins meant the wind had a greater influence on that part of the rocket, thus bringing the center of pressure toward the rear, far from the center of mass, and causing the rocket to be dangerously "overstable." Boom.

BUILDING A BETTER BOARD

Originally, Carbon Origins just wanted to make the best rocket they could. But they realized that that rocket needed a brain that wasn't available yet. So while other space

startups and rocketeers are experimenting primarily with engines and the rockets themselves, to Carbon Origins the rocket is the vehicle — literally and metaphorically, says Shara — by which to launch the Apollo controller. (A mailing list currently allows you to reserve one, and more than 5,000 people have done so. Two versions are planned, entry-level and pro.)

"The electronics is such a huge part of launching a high-power rocket," says Shara. "It's not important in little rockets, but in big rockets it's all about tracking and data and all of the different equipment and electronics and cameras on the rocket."

"It's a very extreme environment, it really gave us an opportunity to build something at a very high level," she continues. "We're building something that can exist and work in an environment that would be probably beyond a lot of people's projects."

The board is built with flanges that draw heat to the gold-plated edges, making Apollo act like its own heat sink and RF shield in one. Sensors are delicate where heat is concerned, so this layout means components can be placed closer together, allowing them to mash 11 sensors onto the

tiny board.

A 32-bit ARM Cortex-M3 processor runs GPS, Wi-Fi, and Bluetooth, as well as accelerometer, magnetometer, pressure sensor, infrared and ultraviolet light sensors, and more. The face of the approximately 1"×2" board is largely covered by an OLED screen and a trackball to toggle through apps and data.

With its tiny size and myriad of sensors, Apollo is unique within rocketry, but it's also unique among boards in general — it has attracted potential users far outside rocketry, including wearables, drones, and Internet of Things devices. The Void, a virtual reality arcade near Salt Lake City, is building a VR gaming environment in a theater. Each gamer wears a vest, helmet, and gloves equipped with sensors based on Apollo to track their motion. DomeCandy Labs, a portable speaker maker, is using Apollo's communication capabilities to prototype a Bluetooth speaker that will offer feedback tailored toward music.

Carbon Origins is so bullish on this thread that they're spinning off Carbon Labs, which will mimic Carbon Origins' approach to democratizing space, but for data sensor hubs — i.e., making it easier to use Apollo and its components to meld the physical and digital worlds and enable the Internet of Things. Carbon Labs will offer custom-built products based on Apollo architecture; that is, they'll work backward from Apollo, picking the relevant parts and software for each client, greatly reducing the time required for development, according to Srirangarajan.

Both aspects, the controller (and its potential use in the Internet of Things revolution) and the rockets themselves, are important to Carbon Origins' mission.

"I can't choose sides," says Srirangarajan. "I spend almost the same amount of time on both, they're like my two babies."

They've made it this far without any outside investment, relying on savings and funds from other successful startups they've been involved with. "Now that we are cash-flow positive, we don't have an immediate requirement," Srirangarajan says. "But it is in our roadmap." Once Apollo launches as a product, they'll explore venture capital.

ORIGIN STORIES

Unlike the Mojave, Case Western Reserve University is not a great place to launch rockets. Still, upon arriving as a freshman, Srirangarajan quickly founded the rocketry club. And a robot club. It was robotics that first hooked him, as a boy living in India, where his mom was a computer engineer who made him learn programming as a kid — BASIC, Java, C. He used to visit a friend of his grandfather's in the country, who had bought a bunch of land after retiring from the Indian navy. Together, they experimented with solar pumps and irrigation. "I used to go there in the summer, and it used to blow my mind, the things you could do with electronics," says Srirangarajan. He built his first robot sometime around age 11, with the help of that friend.

"It's kind of been part of a pretty hard education system, with this insanely cool outlet during the summers. That combination led me to build flame-throwing robots and laser harps and all these really cool Maker projects that not a lot of my peers were really into," he says. That led him to entrepreneurship: "I made stuff, I sold stuff, I made money that helped me make other things."

Robotics led Srirangarajan to space — after all, robots are frequently crucial parts of space missions — and space led him to his co-founders.

Torio, Carbon Origins' COO, was born in the Philippines, and joined up simply because she loves space. "I found out really young that that's what I wanted to do, I wanted to be an astronaut. And that just never left me," she says. "Going the whole NASA route is so tedious, there's no surefire way of getting to what you want to do."

Rocket club was an exception to the rote tedium of PowerPoint classes: "There's this exciting thing that you can actually participate in, actually put the theoretical stuff you've learned into practice, which was amazing."

"College isn't right for everyone. That was definitely the case with me — I hated it," she says. "I'm a mech-e, I graduated as a mech-e, I did not see anything hands-on until senior year." She skipped graduation to go to a robotics competition in Florida.

While Torio skipped graduation, Shara

"WE BUILT APOLLO AS A SWISS ARMY KNIFE FOR MAKERS AND DEVELOPERS."

left school entirely to start Carbon Origins, acting as vice president of electronics. "You don't get great opportunities like this that come along very often," she says. "It was starting to extend beyond the scope of what was possible financially, in terms of resources, and beyond that, acceptable in a university setting."

The decision wasn't out of character for her. She was highly recruited by the engineering school, says dean Jeffrey Duerk, who describes her as "exactly the type of student you want at the university." She quickly became a TA and a leader among her peers. Later on, when her grades flagged, he discussed it with her. "She said, 'I came here to learn, I didn't come here to get good grades.'"

Shara has been immersed in electronics since almost before she could walk, when her father gave her a scrap circuit board. "As soon as I discovered I could find these circuit boards in everything, all the household appliances, our house kind of became the e-waste central of my hometown," she says.

She admires Arduino, and cites it as inspiration for Carbon Origins. "We've come such an incredibly long way, in terms of electronics, that things that are possible or would take financial inputs that only governments can provide, now you can basically order it on Amazon and have it on your doorstep in a couple of days." That's what ties the rockets to the board: the goal of democratizing access to space.

Shara and Dixon both grew up firing little rockets with their dads — she in Montreal, Canada, he in Ann Arbor, Michigan. Dixon couldn't wait to move on to bigger, faster, higher rockets; now he's vice president of aerospace for Carbon Origins. The more advanced he got, the better he liked it. "Once you start getting into harder materials, like aluminum rockets, and making your avionics a lot more advanced, getting data off of them, that's when the real engineering starts," he says. "That's when the problem solving starts. That's what really drives me."

Now they can relive their flights virtually, thanks to the data collected by Apollo. "We like great tools that potentially streamline our engineering, so we built Apollo as a Swiss Army knife for Makers and developers," says Srirangarajan.

One of the best ways to get started
is to check out a local rocketry club. Most cities have them; to find yours, start at nar.org/find-a-local-club.

WE HAVE LIFTOFF

After a second countdown, Phoenix 0.3 achieves ignition. It roars off the launch pad, but somewhere around 3,000 feet the team loses radio contact with it. It goes unstable, spiraling upward in a corkscrew fashion for a few seconds before breaking into pieces near the 10,000 foot mark. The heavier parts tumble down just outside the FAR compound, while the main chute breaks away and floats off to the south.

The Carbon Origins crew wanders out of the bunker, watching the carnage, while Dixon, still on video, asks what happened and where everyone went.

"That's disappointing, I suppose," says Srirangarajan. "But I have a track on a few parts, which means data, and data is good. It's okay, failure is good, because we'll learn."

Some of their simulations showed oscillations a little before Mach 2, Srirangarajan says as he combs the desert for rocket parts, marking the GPS points of each one so he can re-create the debris field on Google Maps. Too much oscillation could have caused the composite body to snap, releasing the parachute, and the drag could have cause the spiral. But that's a guess; to find out, they'll have to locate the flight controllers, which seem to have floated off with the main chute.

Those three Apollos and six commercial flight controllers were housed partially in the nose cone, partially in the body. But when Srirangarajan finds the nose cone, it's empty, embedded tip-first in the sand. The wreckage of the lower half of the body lies not far away, black aluminum fins bent to the side. The electronics bay is there, still beeping, but it contained just a pair of radios and three off-the-shelf flight controllers. "This really gives you an appreciation for the strength, and the forces," says Shara, holding it up.

"Rockets are hard," she says again. "It wasn't, certainly, a complete failure." If they can recover the Apollos, the data will tell them what went wrong. And with luck, they can still compare the data between Apollos, and from the commercial controllers, to see how accurate the board is.

It takes them more than a week to find the main chute. They spend days crossing the desert, search-and-rescue style. They

"THREE SECONDS IS AN ETERNITY IN ROCKETS."

Launch Log
The following infomation is derived from raw data from Apollo and other onboard computers during the launch of Phoenix 0.3. "We have not yet arrived at any conclusions or definitive reasons for what happened," says Srirangarajan.

[Seconds]

[0.00] Liftoff.

[0.25] Rocket cleared the 16' launchpad. It is now traveling at 112mph and accelerating at 17.25g.

[1.04] The rocket starts to roll at 1.2rpm. This is much higher than the expected roll rate of no more than 0.5rpm until engine burnout.

[1.79] Transonic speed achieved. "Mach-delay" initiated. The data from the barometers cannot be used to calculate the altitude after this point. Last recorded "good" barometer data was at 2,250'. The rocket is now accelerating at 20.8g.

[2.08] Unexpected temporary spike in thrust recorded.

[2.15] Supersonic.

[2.16] 50% fuel consumed.

[2.18] Max thrust of the rocket engine is recorded as 1,185lb. This is about 10% higher than expected.

[2.21] Another unexpected temporary spike in thrust.

[2.28] Roll rate now at 20rpm.

[2.35] Highest aerodynamic stability during boost stage reached.

[2.38] The rocket is now accelerating at 21.2g (max-g).

[2.46] Horizontal vibrations exceed 5g range.

[2.75] Mach 1.5.

[3.10] Roll rate starts to decrease. Max roll rate 23.8rpm.

[3.18] Mach 1.75.

[3.25] Wobble (angle of attack exceeds 0.5°).

[3.40] Mach 1.9.

[3.41] Rocket now accelerating at 20.3g.

[3.42] Shred.

[3.43] In tumble. Parts coasting up and decelerating.

[6.79] Mach-delay lift.

[9.30] 9,250' apogee on electronics bay from barometric data. Inertial measurement unit (IMU) data too complex to derive altitude information.

[11.12] 12,500' apogee on nose cone/main parachute computer from barometric data. IMU data too complex to derive altitude information.

rent ATVs and drive around. Finally, they program their Phantom drone to fly in a grid pattern 150 feet above the ground. The red-striped parachute stands out bright against the desert brush, but it had drifted 6 miles southwest from the launch site.

Attached was a green payload board, carrying one Apollo and three commercial flight computers. Apollos talk to each other, notes Srirangarajan, so if they can get data off it, they'll have a successful mission after all. But it's got a cracked screen and the SD card may be damaged; they can't risk ejecting it until they know for sure. Instead, they plug the board in and arm it as if the rocket were still whole. It shows up as an external drive, and they download the CSV file. Combined with high-speed video and the data from the commercially available boards from the electronics bay, this gives them a picture of what happened.

Only 1 second after takeoff, Phoenix 0.3 starts to roll — i.e., rotate around its vertical axis — at more than twice the expected 0.5rpm rate. It goes supersonic, and its records show greater than expected thrust. By 2 seconds in, the roll rate has reached 20rpm. That wouldn't be such a big deal, but combine roll with a little tilt and a little vibration, and you start to spiral.

It hits nearly Mach 2 before it splits apart. The parts decelerate, but continue to shoot upward until about 10,000 feet, when they reach apogee and start to tumble back to Earth. It takes less than 4 seconds from launch to shred, and around 10 seconds to maximum height. "Three seconds is an eternity in rockets," Srirangarajan says later.

By the time they've put this timeline together, Carbon Origins has reached a decision. They're saying goodbye to the house, goodbye to California City, and moving to an actual office in Palmdale, closer to Los Angeles. They'll still launch rockets at the FAR site, still travel together, work together, even live together, but they'll no longer share a living room with a drill press and a shop vac. The new space will allow them to expand, recruit new employees, and move on. They have a two-stage launch scheduled for September, shooting for 180,000 feet. Phoenix 0.3 may not have reached the apex they hoped for, but Carbon Origins is still ascending. ◑

6th ANNUAL NEW YORK

GREATEST
SHOW & TELL
ON EARTH

SEPT 26+27
SAT 10AM–7PM
SUN 10AM–6PM

NEW YORK
HALL OF SCIENCE
QUEENS

WORLD Maker Faire®

NEW YORK
HALL OF SCIENCE

makerfaire.com
Brought to you by Make: magazine

Stuck on Sugru

Inventor Jane Ní Dhulchaointigh talks about finding her product's niche

SUGRU IS A "MOLDABLE GLUE" THAT HAS FOUND A PLACE IN MANY MAKER TOOLBOXES. Born in Ireland, inventor and chief executive Ní Dhulchaointigh developed the first version of Sugru in 2003 while studying at the Royal College of Art in London. After six years of R&D, her product finally launched in December 2009. The first batch sold out in six hours. Sugru now has 45 employees and hundreds of thousands of customers worldwide.

Could you have launched Sugru before the internet?

No way. Simple answer. In the past, for companies like us, the only way to reach customers was through retail. Trying to sell something as complicated as Sugru in a store is a challenge. The package has to explain a lot. I don't think we would have been able to do it.

So Sugru is as much about communications as it is a chemical and product company.

Definitely. There are two sides to Sugru; one is the product, and one is the culture of using it, which we build through communication, whether it's the users producing the communication or us. Without the community and the communication, none of the rest is possible. That's what creates the whole demand for Sugru, and helps us get into the retailers: the brand, the way people are using it, and what they are saying about it.

Did it take you a while to refine that bigger message?

No, that's the whole reason I bothered to do Sugru. I spent five or six years developing the science before we launched. I wouldn't have bothered to do that if I didn't think it was important. I'm only doing it because I really believe the world will be a much better place if people have more confidence about fixing things and making things. We've lost so much in becoming a consumer society. There's this rebellion now that says that's not enough. That's what the Maker Movement is all about. That's why people are so into baking things and growing things. We're all realizing that it's not enough to just buy stuff in shops.

Community has been a huge part of Sugru's success so far.

We never predicted just how vibrant the community would be. Our community has been teaching us what Sugru is good for, and what the problems are out there, which helps us focus on the things that really matter to people. It really helps with product development. That's how we've learned which colors people use, and what package sizes they like, for example.

Giant U.S. retailers, like Target and Lowe's, are now carrying Sugru. What effect has that had on the company?

It's a huge step for us. The customer we're speaking to is much more mainstream. They don't call themselves Makers, but they are keen doers. They're the people in their family who, when something needs fixing, when something goes wrong, they do something about it. They love home improvement. We're getting more focused on that group. Makers are core to what we do, and everyday doers are an extension of this.

What's your advice to other Makers working on potential products?

It's much easier to get user engagement, and get people spreading the word, when you have a common goal and you really want to have a positive impact. Be open about your mission. Communicate it clearly and in a compelling way. Our customers know what we're like. We put ourselves out there. We've done a lot of talks, blogged a lot, made a lot of videos. We're not just trying to build a market, we have a mission: We're trying to make the world a better place. At the end of the day, people like to be part of something. It's important to be genuine. They can believe you're genuine if you put yourself into the picture.

Are there lessons to be learned from duct tape?

Duct tape and super glue are the handiest things ever. That's why they're still here. That's our ambition for Sugru. It's super-inspiring that duct tape has been around since, I think, World War II. It's in everyone's kitchen drawer, and that's where we want to be.

It takes time to achieve that universal status — it's a long game. Another thing is, we have a battle between versatility and simplicity. Versatility can be daunting to new people. What we need to remember is, "keep it simple." People should know what Sugru's good for and discover new uses over time. ●

Dan Dennison

For more Maker Pro news and interviews, visit makezine.com/category/maker-pro, and subscribe to the Maker Pro Newsletter at makezine.com/maker-pro-newsletter.

DC DENISON is the editor of the *Maker Pro Newsletter*, which covers the intersection of Makers and business, and is the former technology editor of *The Boston Globe*.

ILLUSTRATION BY MATTHEW BILLINGTON

THERE'S NOTHING AS DIY AS HOT-RODDING YOUR WHEELS. Our cars are our palettes, representing who we truly are through personalization of everything from the color of the paint jobs to the size of the engines.

We take this customization very seriously. Scores of enthusiasts spend countless hours perched over open hoods, changing fluids, tuning carburetors, and cleaning air filters. The more intrepid overhaul engines in search of better efficiency and, more commonly, faster speeds. And the true diehards build entirely new machines, fabricating bodies and frames to hold the components of their choosing in every imaginable configuration.

We now live in the era of the connected car, with automobiles so sophisticated that they can even think for us. Despite this, we haven't relinquished our right to hack our rides — every automotive advancement and new sensor network has someone pulling it apart to figure out how to make it their own. Join us as we celebrate these trailblazing technicians with all manner of vehicle projects big and small, for everything with one wheel to four. Your garage awaits you.⦾

WORLD'S CUTEST GO-KART

DESIGNING AND BUILDING THE CHIBI-MIKUVAN FOR THE POWER RACING SERIES

WRITTEN BY CHARLES GUAN

CHARLES GUAN is a 2011 MIT Mechanical Engineering graduate, and current machine shop and design instructor. Like many of his peers, he was inspired to pursue robotics as a hobby by the TV show *BattleBots*.

Zachary T. Nguyen

SINCE MY FRESHMAN YEAR AT MIT IN 2007, I'VE BEEN BUILDING SMALL ELECTRIC VEHICLES AS A HOBBY (why *walk* when you can roll in style?). My involvement in the student project space at MIT ultimately spurred me to become a mentor for students building their own projects of all types. Interestingly, the most valuable advice I often gave was not how to make things, but where to get things.

In my years of building vehicles (and robots), I had unknowingly amassed a huge repository of places to buy or scrounge parts, and methods of vetting them for a design. These tactics proved indispensable: as it turns out, nobody teaches you these practical skills in engineering school.

The first time I saw the Power Racing Series (PRS) was at World Maker Faire New York in 2012 — grown adults racing heavily modified or custom-built Power Wheels cars to relive their childhood years. It got me thinking about how to channel my students' efforts into a project that could benefit engineering education.

I began to write down and distill the rag-tag EV building methods that had been refined by students and instructors through the years, and in 2013, I taught the first semester of a lab class which would come to be known as "2.00GoKart" by the undergraduate population. I modeled it after MIT's robotics-competition-focused course ME 2.007, but with students instead building electric go-karts in teams of two.

Students were forced to search for all their own parts, justifying to myself and fellow instructors why their part worked for their design. I ensured that everybody knew what McMaster-Carr was, how to find a certain sized screw there, and why certain decisions came back to bite them later in the semester (like, needing 5 different wrenches to tighten a motor mounting bracket).

IF IT MOVES, USE IT FOR PARTS

With the "2.00GoKart" class increasing in popularity and the 2014 Power Racing Series rule revisions becoming ever more compatible, it was a curious case of convergent evolution. I decided to create a "technology demonstrator" vehicle for PRS 2014. After watching the 2013 race, I thought the sport needed an injection of varied parts and technologies. There were just too many salvaged floor-scrubber motors and forklift motor controllers for my liking.

My intent was simple: show that you

A small "hub motor" scooter I built in 2008 that I used as transport for 2 years.

A uniquely shaped student go-kart runs a time-trial race in the 2013 final contest at MIT.

AN OVERVIEW of the Chibi-Mikuvan, the world's undisputed cutest racing vehicle.

Electric bike throttle

Recumbent trike double brake lever

¹/₅ scale R/C car ESC

Razor go-kart seat

¹/₅ scale R/C boat motor

2.50-4" size hand-truck tires

12V water pump

³/₄" go-kart axle and bearing blocks

9" angle grinder gearbox

POWER SYSTEM

Chibi-Mikuvan comprises a mishmash of parts: an R/C boat motor, R/C car motor controller, angle grinder gearbox, pieces of a hybrid car battery, an ammo can, scooter brakes, wheelbarrow tires, an electric bike's throttle lever, an Ethernet cable, and a USB phone charger. And of course, an Arduino.

My biggest "look for parts in everything" victory came in the form of the battery. In the Power Racing Series, batteries are *not* included — they count against your team budget, and the only way to achieve good runtime has been using lead-acid batteries, which are heavy and inefficient.

I began investigating nickel-metal hydride (NiMH) batteries, which were produced for the first generation of hybrid cars in the mid-2000s. I called no less than 15 auto parts yards and recyclers, and one Friday morning in October 2013, I drove from Boston to Burlington, Vermont to collect a $300 battery from a 2009 Ford Fusion hybrid to disassemble and turn into smaller modules.

The process is *extremely* dangerous, but in the end, I got four batteries for Chibi-Mikuvan out of one Ford Fusion hybrid pack. Ultimately the cost of each battery was only $37.50, and they weighed only 25 pounds.

ANGLE GRINDER GEAR TRAIN

I wanted to experiment with using the latest generation of "giant-scale" R/C model parts, including motor and controller setups, due to their commodity status and low price. R/C motors, however, tend to be designed to spin very quickly with little torque. To tame that power for a ride-on vehicle weighing 100 pounds or more, I'd need a gear ratio of 20:1, compared to the average electric bike/scooter motor used in PRS entries of 5:1 or 6:1.

Creating a gear reduction from bike chain or commercial gears would have been too heavy and large, so I recalled an artifact from my younger years taking apart old power tools: angle grinders. What are they but a motor and a set of hardened steel gears in a premade box?

Mounting the R/C motor to the grinder's gearbox would yield around 4:1, from which I could run a simple 5:1 chain drive using go-kart chain, with the result being the 20:1 needed.

Chibi-Mikuvan on top of 3x scale model of itself.

Chibi-Mikuvan's running gear; all of the control electronics except for the main power switch and fuse are inside the waterproof surplus ammo can.

can use parts from a variety of seemingly unconnected industries, sourced from internet shops and real-life stores alike, to create a highly competitive racing vehicle without a lot of sophisticated manufacturing equipment. The same theme would be useful to my students building practical vehicles, or really any kind of project: that you can use *anything* for parts. Here's the story of the Chibi-Mikuvan.

BODY STYLINGS

You might be thinking: That's not like any Power Wheels car I've ever seen. Chibi-Mikuvan is basically what a Power Wheels version of my car *would* look like — specifically my 1989 Mitsubishi Delica van, an obscure model sold only between 1987 and 1990 in the United States. (Several PRS teams used this new "rule" to great effect in 2014.) To fabricate the body, I chose foam-cored fiberglass construction: lightweight, rigid, and easy to repair. I dedicated the vehicle to Hatsune Miku, a popular Japanese animated musical character.

Zachary T. Nguyen

Charles. Guan

A hybrid car battery module, made of many smaller low-voltage cells which can be reconfigured.

This soda can-sized motor for ⅕ scale R/C boats can output several thousand watts of power — but it spins at over 20,000rpm to do so.

The insides of an inexpensive imported 9" handheld angle grinder's gearbox.

The motor and grinder gearbox mounted together on the frame.

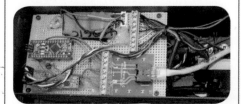

The electrical deck of Chibi-Mikuvan: an Arduino Nano, a modified USB charger as power supply, a high-current relay between battery and motor controller, and a current sensor. The red component is a backup logic power supply.

Pete Prodoehl

Chibi-Mikuvan during a lap at Detroit Maker Faire 2014.

CHASSIS AND BRAKES

The frame is simple welded steel tubing. Hand truck wheels (which didn't last long under racing conditions) came from a local discount tool chain. The rear axle is supported in commercial steel bearing blocks. The only custom parts here are the massive 7" front disc brakes, for which I used an abrasive water jet cutter to make the hub and rotor, pairing them with cable scooter brake calipers. The inside joke is that Chibi-Mikuvan can stop not only itself, but the three karts behind it.

CONTROL ELECTRONICS

The interfaces between driver and motor controller live in a surplus ammo can. It's waterproof, highly shock resistant, and easy to mount. I reused a high-current relay from the hybrid battery as a remote start; turning on the controller is a matter of pressing a small button on the handlebar. An Arduino Nano takes the analog throttle signal and converts it to R/C servo pulses for the motor controller. Since Arduinos can't be powered straight from the 28-volt battery, I repurposed a cheap car accessory, a lighter plug USB charger, as a DC/DC converter.

Chibi-Mikuvan's hodgepodge powertrain proved extremely quick, even if my proto-typed electronics were unreliable — in 2014 it won a sprint race at Maker Faire Detroit and the quickest qualifying lap time at World Maker Faire New York. My hope wasn't to make the absolute fastest PRS car, but the most documented and repeatable one, which can be used as a resource for anyone else. Complete build posts and bill of materials are on my blog at etotheipiplusone.net/?page_id=3434. ◐

Zachary T. Nguyen

See more photos and read the extended version of this article online at makezine.com/go/chibi-mikuvan.

MAKING THE SWITCH

WRITTEN BY PETER OLIVER AND JIM MCGREEN

BUILD THE SWITCH ELECTRIC VEHICLE — A MAKER-READY KIT AND EDUCATION SYSTEM

Time Required:
A Week
Cost:
$14,000–$30,000

Switch Vehicles founders PETER OLIVER and JIM MCGREEN each have a long EV history. Peter taught college students how to convert gas cars to electric, and converted many classic vehicles with his company Make Mine Electric. Jim started Electrathon racing in the early '90s and founded ZAP (Zero Air Pollution), the first electric bike company in 1991. Jim also invented the Zappy micro scooter which sold more than 30,000 units.

Sam Euston

GET SWITCHED ON
Specs and Options

» **DIY kit:** $14,000–$30,000 complete, depending on performance and range options; $6,950 without electrical components
» **Sporty rack-and-pinion steering;** center, left-, or right-hand driver position
» **Accepts wide range of DC and AC motors**
» **3-wheel disc brakes,** independent front and rear
» **Regenerative braking** (with AC motor option)
» **Batteries:** Lead-acid or lithium-ion, 10kWh–30kWh
» **Range:** 45–135 miles (depending on battery option)
» **Charge time:** 2.25 hours per 45 miles, for about $1 (220V 30A power outlet)
» **Gas mileage equivalent:** 51mpge
» **Top speed:** Over 100mph (with AC motor option)
» **Acceleration:** 0–60mph in 9 seconds (or faster with gearing options)
» **Turn radius:** 38' or less
» **Weight:** 1,350lbs (3-seater option)
» **Chain drive standard, belt drive optional**
» **Street legal,** registered as a motorcycle — but doesn't require helmet or motorcycle driver's license!

WHAT STARTED AS AN IDEA TO PRODUCE OUR OWN SIMPLE, FUN, AFFORDABLE ELECTRIC vehicle has evolved into a DIY project easily assembled by nearly anyone. The Switch defines a new class of auto — a lightweight, high-capacity, versatile electric vehicle (EV) that requires minimal resources to construct, assemble, and drive. It can be built in a single-car garage on jack stands, or even in a classroom — all you need is a small box of tools and a 6½-foot door, so it can drive out.

THE SWITCH LAB

It all starts in The Switch Lab — a complete DIY kit and education system consisting of workshops, written and video instructions, detailed plans, and all the parts. This all-inclusive package enables individuals, schools, or businesses to assemble their own version of the EV. (Skilled builders can go it alone with just a DIY kit.)

The Switch Lab workshops help educators and builders gain the confidence and understanding to complete the project. Teachers appreciate how our workshops break a complex technology into clear step-by-step modules. They're also pleased with the optional curriculum — a classroom and lab in a box. "It's really plug and play," said Roger Pressley of Central Academy of Technology and Arts in Monroe, North Carolina. "Any school can do this, regardless of the tools or shop facility. You can start with any space."

Students seem invigorated by the course, and some are asking if they can build the Switch as a business. Whether they become Makers, welders, or engineers, the hands-on learning gives them lifelong lessons, building self-confidence and real-world skills. "It's not just a high school class, it's something you take with you," commented a student from Venture Academy in California.

THE DRIVING EXPERIENCE

The Switch drives like a Formula One car, hugging the ground and protecting you with a NASCAR-

3-phase AC motor attached to swing arm.

Under the dash: dual brake pedals, steering linkages, and accelerator — step on it, but don't call it a gas pedal.

style roll cage. The open cockpit provides an exhilarating experience and excellent visibility for safety. Separate front and rear braking systems enable advanced sport driving techniques.

And, incredibly, you can build it in a week! The chassis is laser-cut, welded tubular steel, and the wiring is professionally terminated, numbered, and color-coded — all ready to be assembled.

CUSTOMIZING THE SWITCH

Imagine bolting on a pickup bed, camper shell, or racks. Select a custom seating arrangement from one to four. Choose from various DC or AC motors for performance options defining power, speed, and regenerative braking. Pick your battery system — budget lead-acid batteries or lithium-ion, from 10kWh to 30kWh battery packs. Adjust the suspension for low, road-hugging traction, or set it high and firm for hauling loads. There are so many options, and you get to decide.

DRIVING INTO THE FUTURE

We imagine Makers taking the Switch Lab process into global markets via micro-manufacturing, tailored to local needs and fueled by regional green power sources. The manufacturing model deepens as regional economies develop, starting with a fully welded chassis kit and expanding into welding or laser-cutting parts to spec as the market grows.

The Switch is designed to move people into another world, a world of voluntary simplicity, respect for the environment and the needs of others. It helps start the transition conversation.

And it's truly a Maker vehicle, ready for the full range of builders. Computer programmers, carpenters, men, women, high school and college students — even auto mechanics — have successfully assembled and understand the Switch. There's no other EV that starts as an affordable project and becomes a sporty utility or off-road transport that you make yourself. ◉

For more information about the Switch EV and the Switch Lab, visit theswitchlab.com.

In January 2015, 12 teachers completed training at our facility, where they assembled two vehicles in one week.

Peter Oliver

EVOLUTIONARY EVS
WRITTEN BY MIKE SENESE
THESE DIY CAR MAKERS ARE PUSHING PAST THE ELECTRIC-VEHICLE BIG BOYS

Jen Danzinger

SEVEN BY ILLUMINATI MOTOR WORKS

This gull-winged beauty with rear-facing back seats has dramatic lines that evoke the classics from the '30s. But the retro vibe ends there; under the hood it's all 21st-century tech. A total of 99 lithium iron phosphate batteries supply 33kWh to a 200hp MES DEA electric motor, pushing it to 130mph and doing 0–60 in just 6.2 seconds. Despite weighing a beefy 2,900 pounds, it can go 200 miles on a full charge. Along with regenerative braking, a 220V 40A charger will replenish it overnight, and the team is even testing out linear generators to recover vibrational energy. And that body? Carbon fiber and Kevlar.

Built by hand in an Illinois barn by a team of alternative-fuel vehicle consultants to compete for the $5 million 2010 Progressive X Prize, the Seven blasted past the 100mpg-equivalent criteria by getting 207.5mpge. Since then, refinements are ongoing — team leader Kevin Smith explains "It's all about efficiency here at the IMW headquarters."

Simone Spada

TABBY EVO BY OSVEHICLE

The OSVehicle team aims to localize car production with its open-source plans — the blueprints and design files are freely available — for a multipurpose car platform that can be easily customized for individuals' and groups' specific needs.

Their latest iteration, named Tabby EVO, steps up from their original 2013 release with refinements to the assembly process and to the car's driveability. It features an 80V/15kW electric drivetrain with a range of 87 miles and a top speed of 80mph. It has a 93" wheelbase and can be configured for 2–4 passengers. Its upgraded frame and suspension help qualify it for L6e, L7e, and M1 licensing for street-legal use in Europe and the U.S., as well as allow it to be used for off-road purposes.

OWN YOUR CAR

WILL A VAGUE COPYRIGHT LAW AFFECT THE SCOPE OF WHAT HOBBYISTS AND MODDERS CAN DO WITH THEIR VEHICLES?

WRITTEN BY BENJAMIN PRESTON

BENJAMIN PRESTON is one of the few journalists to hold the dubious distinction of having been an alignment technician at a Pep Boys in Fredericksburg, Virginia, as well as an automotive reporter for the *New York Times*. In addition to the *Times*, he has written for the *Guardian*, *BBC Autos*, *Car and Driver*, *Jalopnik*, and, he is very proud to say, *Petersen's 4-Wheel & Off-Road*.

THE SUBURBS OUTSIDE WASHINGTON, D.C., MAY BE BETTER KNOWN FOR THE HIGH TECH TELECOMMUNICATIONS AND defense industries than for automotive hot rodding, but in a modest two-bay shop in Northern Virginia, Michael Hogarty is one of the many players eager to upset that notion. All around the barn-like structure where Hogarty runs his professional repair shop are hints of an atypical operation. A Suzuki Samurai with a more powerful Chevrolet motor installed sits in one corner of the driveway. Evidence of a wide range of projects — custom-made vehicle accessories and skateboard ramp components among them — lean against the fence.

Michael Hogarty

Hogarty, an ASE-certified master technician, began his career working in other people's shops, fixing lawn mowers and later, Ferraris. Now he spends a good portion of each week diagnosing and repairing mundane family haulers. But what he's really into, he says, is custom modification, which is, at its root, problem-solving. If a customer wants a new stereo, a nonstandard engine, or wheels that don't fit, he finds a way to make it work, and work well. His shop is abuzz every day of the week with power junkie friends and customers who want setups no one else knows how to do.

But with vehicle computer systems becoming ever more complex, even jobs that were once simple, like replacing the stereo, are much more involved than they used to be. His projects have him digging ever deeper into the computer code that runs vehicle systems in an effort to balance the changes made when individual components are swapped out or improved upon.

"On newer vehicles, you can't just pull out the stereo and throw in something from Best Buy," he says. "Vehicle systems are all integrated now."

So he relies upon a broad knowledge of mechanics, electronics, and computers to find the best outcome for every puzzle that comes his way.

But there's a challenge on the horizon different from the custom turbo installations and remote starter setups he has tackled in the past. It's a legal hurdle, but it's also a cultural question. Hogarty, and other independent mechanics and innovators like him, may see an obscure copyright law passed nearly two decades ago — the Digital Millennium Copyright Act (DMCA) — impact their ability to access the computer systems central to today's cars and trucks. Modders have become more like hackers than ever before, and the boundaries of how they can exercise their computer savvy is at stake.

PROGRAMMING PERFORMANCE

The brain of every modern vehicle is its computer system. Like a brain — or brains, since the system is really a network of computer modules working in concert — the system sends and receives signals to and from sensors located all over the vehicle through the Controller Area Network, or CAN. Among its primary functions is engine management, the delicate balance of supplying fuel and air in the right doses to create smooth, efficient power. But it also regulates heating and cooling, audio, braking, stability, air bags, door and window motors, and other things. CAN data allows different systems to communicate with one another, and makes possible features like speed-sensitive stereo volume control and automatic sliding doors that cannot be opened while the vehicle is in motion

Vehicle computers are nothing new. The electronic control unit (ECU) as we know it today appeared in the late 1970s. By the early 1980s, all cars in the United States were equipped with decision-making microprocessors. The earliest used input from engine sensors to activate solenoids and actuators designed to optimize engine performance. General Motors, for instance, used computer-controlled carburetors that reacted better than mechanical ones to changes in atmospheric conditions. Over the years, other functions were gradually added to the list of tasks ECUs could handle. Wire looms became ever more bulky, so the system was streamlined with the advent of the CAN (rather than giving every system its own wires).

Through the CAN, and specifically through the onboard diagnostic, or OBD-II port (a

Juliann Brown and Hep Svadja

Andrew Albosta

MODDERS HAVE BECOME MORE LIKE HACKERS THAN EVER BEFORE

multi-pin computer connector that's been installed in all cars in the U.S. since 1996, within a few feet of the steering wheel), Hogarty and his tinkerer cohorts can access vehicle software to find patterns and make changes when necessary. Each vehicle may have been engineered by the manufacturer to satisfy emissions, safety, and fuel efficiency requirements, Hogarty says, but someone interested in modifying one may have more specific goals. Many want more power for better results at the track, but lower emissions, better fuel economy, improved cooling for towing, and more complete engine performance monitoring are among myriad other reasons people have for accessing and changing factory settings through a vehicle's on-board software.

Paul Bartek, an engineer for a company that builds electronic displays, dabbles in the auto hobby in his spare time. But his tinkering goes far beyond the mechanical trial and error of years past. Through his website, cowfishstudios.com, he has shared open-source software, tutorials, and DIY projects aimed at automotive enthusiasts. One uses a Raspberry Pi to extract engine performance numbers and display them on the vehicle's display screen.

"I created this because I wasn't happy with what was on the market," he explains. "There were a few apps that could show engine torque and a few other things, but there's a lot more information available. There was no user input on any of those apps."

Bartek isn't alone. There are all sorts of products on the market, many of which had the same humble beginnings. One, called Clickdrive, integrates driving-related smartphone apps. Another, Truvolo, monitors fuel economy and location. There's another that keeps tabs on traffic conditions, coordinates ridesharing, and helps users find parking and filling stations. Hogarty says he uses aftermarket piggyback connectors that tap into vehicle control computers so that he can change parameters like transmission shift points and temperature settings.

Progressive Insurance offers a device called Snapshot that beams driving habits to insurance adjusters from your OBD-II port through a cellular modem. The company gives a discount to customers who drive softly, and for as few miles as possible — what it calls low-risk driving. (Of course, there's a hack for Snapshot that allows the user to trick the device into thinking it's plugged into the car of a little old lady who only drives to church every Sunday.) Other apps are aimed at parents who want to make sure their teen drivers aren't distracted while behind the wheel.

Like Bartek, Hogarty uses small microprocessors to change values in the data that engine sensors send to the computer to optimize power. For example, swapping the air filter housing from the restrictive closed box most cars get at the factory for an open element filter changes the engine's airflow characteristics. During warm up, and at wide-open throttle, when the most demand is being placed upon the engine, vehicle computers operate in "open loop" mode, using factory-set values to determine fuel flow. Hogarty says that by installing a small microprocessor in-line between the mass airflow sensor and the computer — feeding it power from nearby wires, as well as signals from other sensors, like the throttle position and O_2 sensors and the tachometer — he can trick the computer into changing those values at wide-open throttle, dialing in air fuel requirements for more specific conditions.

"I see the input, measure the output, then change it to what I want it to be," he says. "The manufacturer spends an incredible amount of time tuning their software in certain areas, but not at full load. They build their cars to satisfy a wide range of conditions, but that's not necessarily going to work out when your goal is to squeeze as much power as possible out of it at the track or something."

Hogarty says he has also used micro-processors and data analysis to help customers with more mundane problems. By installing a $30 module — something like a Raspberry Pi — and some open-source software to a vehicle's computer system, a customer can, if they notice that the "check engine" light has come on, text him engine data from a specified time period. It helps him diagnose the problem and saves everyone a lot of time.

COMPUTERIZATION CREATES QUESTIONS

Could the DMCA interfere with projects like Bartek's? Could it keep Hogarty from diving into a computer system to change performance characteristics? Would its language prevent the people who developed other apps and programs from creating more? The law includes a provision regarding "technological protection measures (TPMs) that control access to copyrighted works." The Electronic Frontier Foundation submitted a petition to the copyright office earlier this year proposing an exemption that would make an end run around the TPMs. General Motors opposed the exemption, calling it "overbroad," and said in its comments that proponents "failed to establish that the challenged TPMs are causing, or are likely to cause in the next three years, a substantial adverse impact on users."

The question is, whose side of the story is the right one: the auto manufacturers' or EFF's? According to the EFF, each state's court system can interpret the law differently, which could lead to overzealous enforcement by automakers.

"At best, you're left in a situation where manufacturers can threaten people who access vehicle software, which has a chilling effect on research and tinkering and means that innovators find it difficult to raise funding for new products that make cars more useful," Kit Walsh, a staff attorney for EFF, says in an email. "A clear rule from the Library of Congress is needed to fix this state of affairs and protect legitimate tinkering and research in practice."

Daniel Gage, a spokesman for the Alliance of Automotive Manufacturers — an advocacy group that includes all three major American automakers, as well as most of the Japanese and European ones — says in an email that the information protected by the DMCA was not needed to diagnose and repair vehicles.

"Automakers are concerned that the proposed changes to the DMCA would cause dangerous consequences, including additional safety risks and violations of existing safety and environmental laws," he says. "I also am not aware of any widespread DMCA prosecutions as a result of auto activity now, so what's legal and acceptable today should be tomorrow unless changes to the law are made."

EFF counters that making sure federal emissions and safety rules are followed is up to agencies like the Environmental Protection Agency and the National Highway Traffic Safety Administration, not manufacturers and copyright law.

"Copyright law has crept into this space as a result of computerization," Walsh says. "Under the manufacturers' theory, this upsets traditional concepts of ownership that arise when you purchase a vehicle and lets them prohibit modding that involves software."

Hogarty at the console of a dynamometer, which allows him to verify how changes to an ECU affect torque and horsepower.

YOUR CAR, YOUR CHOICE

In his backyard shop, Hogarty is confronted by a problem with a Toyota Sienna minivan that's a few years old. The sliding door won't close all the way on cold days, or when the vehicle is parked on a hill. It's an electronic malfunction, and to repair it, he says he has a couple of options. He can replace the motor, which still works, but that would be costly. So he taps into the vehicle's computer and changes an electronic threshold value for the door so that it closes, even if the motor has to work a little harder and draws more current. For the family who owns an older van, it's a better option.

He's concerned that without more resolute language from the Copyright Office concerning DMCA, his solution to that problem and his deep dives into engine data to improve performance for motorsports applications could evaporate. "I think it's best to reduce government and corporate input into our affairs and let the individual assume liability for the modifications made," he says.

If the recent spate of articles charging automakers with stripping customers of full vehicle ownership is any indication, the law's vague language has plenty of people worried. As anyone who has ever worked on a car before knows, mistakes and unintended consequences are inevitable, but part of the learning process.

"Do we, as a country, want to have the ability to make mistakes on the things we purchase?" Hogarty says. "I generally don't recommend that people go too far into the back end, but if you're messing with your own car, like anything else — it's up to you." ⊘

Hogarty tests changes to the ECU of an SUV as it sits on a dynamometer.

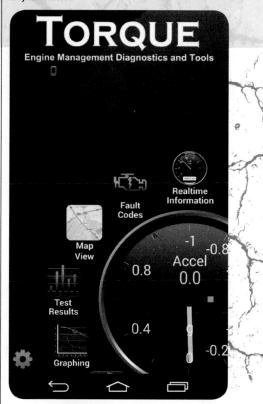

The Torque app for Android accesses your car's OBD-II port to track engine data and display it on your smartphone.

MegaSquirt, a DIY electronic fuel injection computer.

MAKER'S DASHBOARD

DON'T LET THE LIMITED FEATURES OF YOUR CAR'S DASHBOARD GET YOU DOWN.
HERE ARE SOME PROJECTS THAT WILL HELP YOU CUSTOMIZE YOUR COCKPIT.

WRITTEN BY DONALD BELL • ILLUSTRATED BY DAMIEN SCOGIN

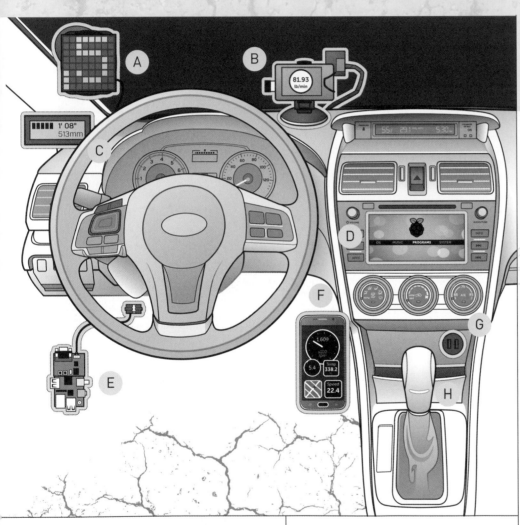

D. IN-CAR ENTERTAINMENT SYSTEM
MATERIALS: Raspberry Pi + 7" TFT monitor + wireless keyboard
COST: $100
Use a Raspberry Pi and monitor to put an in-car entertainment system in your dashboard, or entertain passengers in the backseat.
makezine.com/go/pi-dash

E. SNIFF OUT CONTROL AREA NETWORK CODES
MATERIALS: Raspberry Pi + PiCAN board + OBD-II shell and pins
COST: $100
Beyond your car's basic diagnostic data is a CAN system that controls everything from the buttons on your steering wheel, to your wind-shield wipers. Use a Raspberry Pi to glimpse the code that connects all the elements of their vehicle's electrical systems.
makezine.com/go/pi-can

F. CREATE A CUSTOM OBD-II APP INTERFACE
MATERIALS: Smartphone or tablet + OBD-II Bluetooth adapter + app
COST: $30 (excluding smartphone)
OBD-II compatible apps for smartphones and tablets not only offer a convenient way to tap into your car's data, they're also highly custom-izable and can take advantage of mobile hardware features such as GPS for speed mapping.
makezine.com/go/obd-ii-app-interface

G. DASH-MOUNTED USB OUTLETS
MATERIALS: Panel-mount USB charge port (12v)
COST: $25
Tidy up the tangle of charging adapters in your console by installing a dedicated pair of USB outlets.
makezine.com/go/car-usb

H. WIRELESS PHONE CHARGER
MATERIALS: Wireless charger + 12v to 5v adapter
COST: $50
If your phone supports wireless charging, do away with cable clutter altogether by embedding an inductive charging pad in your console.
makezine.com/go/car-wireless-charge

A. LED MATRIX TACHOMETER
MATERIALS: Arduino + OBD-II TTL adapter + LED matrix
COST: $100
Monitor your engine's RPM and improve your manual shifting precision with a breakout LED tachometer.
makezine.com/go/led-tach

B. CAR DIAGNOSTIC DISPLAY
MATERIALS: Raspberry Pi + TFT shield + OBD-II Bluetooth adapter + Bluetooth 4.0 adapter
COST: $130
Get an inside look at real-time technical data

from your vehicle by transmitting it from your car's OBD-II port to a dedicated dash display and a Raspberry Pi.
makezine.com/go/diagnostic-dash-display

C. PARKING OBSTACLE SENSOR
MATERIALS: Arduino + ultrasonic sensor + LCD screen
COST: $80
Update an older car with an Arduino-powered ultrasonic parking sensor that provides both visual and audio cues to gauge the distance of objects from your bumper.
makezine.com/go/arduino-parking

BIKE HACKS
PERSONALIZE YOUR RIDE WITH THESE CLEVER PEDAL PROJECTS

1. PEDAL POWER PHONE CHARGER
Give your USB gadgets some extra juice while you break a sweat.
makezine.com/go/bike-phone-charger

2. SPOKE AMPLIFIER TAILPIPE
Ever stick a baseball card in your spokes? Now they'll really hear you coming with this stylish bike spoke resonator pipe.
makezine.com/go/spoke-resonator

3. WIRELESS E-BIKE BATTERY MONITOR
Keep an eye on how much juice is left in your e-bike with this wireless Bluetooth LE monitor.
makezine.com/go/ebike-monitor

4. LEATHER BELT HANDLEBAR GRIPS
Easily turn an old belt into a comfortable and classy pair of handlebar grips.
makezine.com/go/belt-bike-grips

5. BIKE SEAT SECRET COMPARTMENT
Prep your bike for the zombie apocalypse (or an off road trek) by hiding an emergency survival kit in your bike seat.
makezine.com/go/bike-compartment

6. TAKE-ALONG BAR
Bring your refined taste in cocktails on the road by building this mini liquor cabinet that fits in your bike frame.
makezine.com/go/bike-bar

7. FRIENDLY FACE TIRE CAP
Pull the head off a Playmobil figurine and transform it into a kid-approved cap.
makezine.com/go/play-cap

8. GIANT STREET CHALK GRINDER
Expand the scale of your sidewalk chalk art pieces by attaching a giant homemade chalk brick to your back tire.
makezine.com/go/chalk-grinder

1. Sean Ragan 2. John Edgar Park 3. Alasdair Allan 4. Nikos Mavrivakis
5. M3G 6. Lee Swenson 7. Urs Graedel 8. Lled Smith

ELECTRO-FY YOUR BICYCLE

WRITTEN BY PARKER JARDINE

FINALLY, AN E-BIKE CONVERSION YOU CAN DO

Extra-wide handlebars have space for e-bike controls and LCD display

Time Required:
1–2 Weekends

Cost:
$1,000–$1,200
Not including miscellaneous bike parts

MATERIALS

» **Electric mid-drive motor, 750W**
Bafang BBS02 kit with LCD display, 25A controller, and thumb throttle

» **Batteries, prismatic lithium iron phosphate (LiFePO₄), 12V 20Ah (4)**
16 cells total. I used #LF-GB4S20 from AA Portable Power, batteryspace.com.

» **Smart LED balance boards (16)**
AA Portable #PCM-BL20CH

» **Series connector bars (2)**
AA Portable #TAB-EL20

» **Smart battery charger, 51.2V LiFePO₄**
AA Portable #CH-LF48V6-TSL

» **Fuse, 30A, and inline holder**

» **Hookup wire**

» **Crimp-on butt connectors**

» **Insulated ring connectors**

» **Heat-shrink tubing**

» **Cable ties, zip ties, spiral cable wrap, Velcro wrap**

MISCELLANEOUS BIKE PARTS

» **Rear bike rack**

» **Rear pannier bags, waterproof (2)**

» **Kickstand** Amazon #B00B29EWPW

» **Lights and horn, 48V (optional)**
Golden Motor #ACC-008 and ACC-002, with combo switch Amazon #B00DGW71D8

TOOLS

» **Hex wrench set**

» **Wrench, adjustable**

» **Wrenches to fit battery terminals**

» **Pliers, crimping** for wiring connectors

» **Heat gun or hair dryer** for heat-shrink

SPECIALTY BIKE TOOLS

These will vary depending on what components are used on your bike. Ask your bike shop!

» **Universal crank puller** for square taper and splined cranks

» **Bottom bracket removal tool**

» **Mid-drive installation tool, or fixed-gear lockring wrench, or Shimano TL-SR21 chain whip** to tighten the mid-drive M33 nut

ONE DAY I BECAME FED UP WITH ALL MY EXCUSES FOR DRIVING TO WORK EVERY DAY. During the summer I would ride my bike, but I was never consistent. There's one huge hill climb that is quite daunting, and is the main reason most people take a vehicle. Finally I said, screw it — I am going to put together an electric bike.

PARKER JARDINE is a manager of systems administration at Fort Lewis College, and enjoys biking, climbing, hockey, camping, mountaineering, hunting, paragliding, and just being outdoors. His tutorials on DIY solar panels and solar systems design can be seen in *Make:* Volumes 12 and 14.

ANATOMY OF AN E-BIKE

Tired of recharging your lights? Wire them to the main battery pack too

The bike literally helped transform my life. I no longer drive to work everyday, I ride instead

See more panniers reviewed on page 89

A mid-drive motor mounted near the crank leaves your rear derailleur free to shift gears

A previous version used a rack-top battery box, but panniers keep the load more stable

I usually charge the battery pack once a week

The kickstand is important for a bike of this weight, so choose a durable option that can support those batteries

Scott D. W. Smith

MY GOAL WAS SIMPLE: BUILD AN ELECTRIC BIKE THAT EVERYONE CAN REPLICATE with minimal cost and without electrical know-how, that's powerful and reliable. I also wanted to decrease complexity over other builds, which would in turn make maintenance easier. This is that bike.

1. CHOOSE A MOTOR

In my opinion, the first item you should decide on is the electric motor. Traditionally, the in-wheel *hub motor* has been most popular, either geared or direct drive.

But I went with a relative newcomer, a *mid-drive motor*, which is not installed in your front or rear wheel, but attached to your front drive system (Figure Ⓐ). The main advantage to this is the ability to use your own bike's rear derailleur to change gears. You can sense the motor's optimum RPM speed, and simply change gears in the rear to get the perfect combination. Another advantage is, with this kit, the bulk of the work is in setting up the battery pack — the control levers are simple to install.

2. (MOST) ANY BIKE WILL DO

The beauty of this electric bike is that you can install the kit on pretty much any bike with a standard 68mm bottom bracket shell width. Choosing the bike to convert is really fun — you can use an existing bike to save a bit of money, or check your local bike shop or Craigslist. Things to look for:

» Strong wheels and fat tires
» Wide handlebars for mounting controls
» Eyelets for rear rack attachment
» Front disc brakes for steep hills

3. INSTALL THE MOTOR

Remove your chain, crank set, and bottom bracket. Depending on your bike, these are nontrivial techniques, so if you're not comfortable or don't have the correct tools, it's best to consult an experienced mechanic or a bike shop.

Slide the mid-drive kit through your bottom bracket and screw it down tight on the other side. The rest is pretty easy. Simply route the cables coming out of the motor to the designated locations on your bike. Then reassemble, and mount the throttle, controller, screen, and brake levers to your handlebars.

4. WIRE AND MOUNT THE BATTERIES

The bike that I decided to use already had a setup for a rear rack. I would simply put battery packs into panniers on each side to distribute the weight. This design keeps the weight down low, making the bike more stable. In addition, the bags can be used to carry extra gear and items around town.

I chose prismatic LiFePO$_4$ batteries, which are often used for solar power storage, for the simplicity of balancing the cells and the ability to remove or replace without having to rip apart the battery pack.

This modular functionality is very important. I began with four 12-volt packs and 16 LED balancing modules (Figure **B**). Using the balancing cells, a smart charger, and the motor cutoff thresholds, you shouldn't need a battery management system. A 52-volt smart charger allows you to charge all the batteries when wired in series, and cuts power automatically when they're full.

Connect the 4 packs in series to quickly test the smart charger, then install an LED balancing module on each of the 16 cells.

Connect a pair of 12V packs in series to make a 24V pack, and repeat for the other pair. Run a series wire from one pannier to the other to connect the 24V packs. If the connector bars don't reach, flatten them with a hammer to extend their length.

Add a 30-amp inline fuse to the positive battery terminal of the battery pack. And make sure to use heat-shrink tubing to insulate each connection.

Finally, measure, cut, and install the positive and negative cables from the panniers to the motor's XT90 connector (Figure **C**). Congratulations, you've electrified your bike — and cut your carbon emissions 90%. ◑

Get more photos, videos, and tips in the extended version of this article online at makezine.com/go/diy-electric-bicycle.

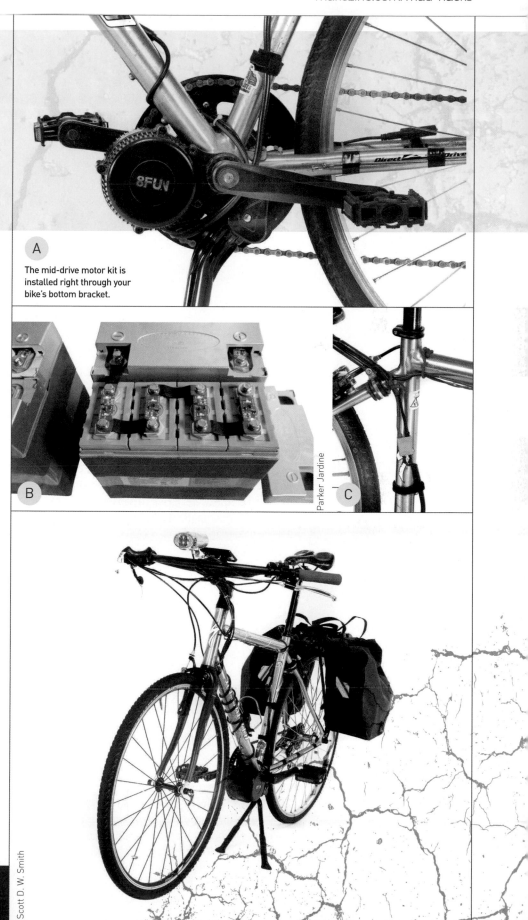

A

The mid-drive motor kit is installed right through your bike's bottom bracket.

B

C

Parker Jardine

Scott D. W. Smith

BOTTLE CAP BIKE LIGHT

UPCYCLE A SODA BOTTLE INTO A LONG-LASTING, AUTOMATIC FLASHER

WRITTEN BY NIKOS MAVRIVAKIS

Time Required: 1–2 Hours
Cost: $5–$10

NIKOS MAVRIVAKIS is a bicyclist and bike tourist from Greece who uses recycled and found materials to make sustainable bike gear. His blog, Bicyclosis, features stories and inventions from his travels. bicycleobsession.wordpress.com

Nikos Mavrivakis

MATERIALS

» **Plastic soda bottles, 20oz, with caps** (2)
» **Rubber band**
» **LED, 1W**
» **Resistor, 10Ω**
» **Coin-cell battery** CR2032, CR2025, or similar
» **Wire, bare or solid-core, 10"** or so
» **Speaker wire, 2-conductor, 5'** or so
» **Soda can, aluminum**
» **Bicycle inner tube, discarded**
» **Reed switch**
» **Small magnet, at least 8mm×2mm** You can find these in old headphones.
» **Zip ties, scrap wood, and rustproof wire** for mounting the light

TOOLS

» **Utility knife**
» **Drill**
» **Sandpaper or high-speed rotary tool**
» **Pliers, needlenose**
» **Soldering iron** (optional)

A

B

C

THIS AUTOMATIC LIGHT FROM RECYCLED PARTS IS ABOUT THE MOST ECO-FRIENDLY DIY bike light in the world. Encapsulated in plastic bottle caps, an LED is activated when a magnet mounted on the wheel passes a sensor, similar to how a bicycle computer works. Anyone can make it — no soldering required.

The bright 1-watt LED will shine through the cap, so choose an orange or red cap for a taillight. If your bike has 26" wheels, it will flash once every 82" you travel. And it lasts up to 320 hours — about 10 times longer than store-bought lights.

1. MAKE THE CASING

Cut the neck and cap off a soda bottle. Trim the neck so it will fit into a second bottle cap. Cut a small rubber band to fit in the channel of the second cap, with a little overlap. This will make it impossible for rain to seep in.

Now fit the neck tightly into the second cap (Figure Ⓐ), then cut off the plastic ring and sand off any markings from the caps.

2. MAKE THE INTERNAL FRAME

Cut 2 strips of bottle plastic slightly longer than the cap's width. Slot the strips together and squeeze into a circular shape (Figure Ⓑ).

3. MOUNT THE LED

Cut another strip of plastic and slide it crossways through the first two. Cut a hole in the center and mount the LED. Connect the resistor to the LED's negative lead, and a short wire to its positive lead. You can solder or just twist these connections.

4. MOUNT THE BATTERY

From the aluminum can, cut a disc the size of your caps and sand off the plastic coating. Drill 4 tiny holes and weave a short bare wire through them as shown in Figure Ⓒ, then connect it to the resistor. Make an identical disc and connect this one to a short insulated wire. Sandwich the coin cell between the aluminum discs, then pack it down snug with a rubber disc cut from an inner tube. Touch the 2 wires together — the LED should light up.

5. CLOSE IT UP

Connect the 2 wires to the speaker cable. Drill a hole in the first cap, pass the cable through, and wrap it with a strip of inner tube to seal the hole.

6. CONNECT THE REED SWITCH AND MAGNET

Connect the cable to a reed switch on your bike frame, and mount the magnet on your wheel. Now whenever the wheel turns, your light will flash! ✇

For mounting tips, video, and complete step-by-step photos, visit the project page at makezine.com/go/waterproof-bike-light.

INSIDE THE MONOWHEEL
FABRICATE THIS FUN AND UNIQUE ROLLING-HOOP RIDE **WRITTEN BY DR. DAVE SOUTHALL**

MONOWHEELS, BY THEIR VERY NATURE, ARE FAIRLY SIMPLE MACHINES; AFTER ALL THEY ONLY HAVE ONE WHEEL! They work by driving a seated base around the inside of a circular track to push it forward — you might think of them as a perpetual roller coaster.

A complete monowheel is usually a custom build. Want to make your own? Here are the key elements:

1. OUTER FRAME

Needs to be big and sturdy enough to sit in, and allow the drive to make it move. I used a 50mm diameter steel tube with a wall thickness of about 3.5mm, rolled into a 1.5m hoop. For tread I cut up bicycle tires and pop-riveted them on the outside of the ring.

2. ROLLERS

These allow the outer ring to revolve around the inner frame. I used custom-made 100mm nylon rollers with ball races to allow free movement. Mine has four, some designs use more. If cost is an issue, you can use skateboard wheels set at an angle.

3. INNER FRAME

Holds the rollers, power source, drive, and seat. Experiment with positions, but keep the center of mass as low as possible for stability.

4. POWER SOURCE

This could be a petrol engine, an electric motor, pedals, or even steam power. It's down to personal preference and engineering ability.

5. DRIVE

Usually a friction-type drive. I use the wheel from a MiniMoto, running against the inside of the outer ring.

Other things to consider in the design: The gearing needs to be calculated; I suggest a top speed of around 10mph if it's your first attempt. Brakes are definitely an issue too — if your braking system effectively locks the inner frame to the outer ring when applied, you will roll completely around with the outer ring, known as *gerbiling*. Best to avoid if possible.

Every monowheel is a unique machine — these aren't rules, just suggestions based on my builds and experience. For more details and ideas go to redmaxmonowheel.co.uk. Enjoy! ◗

Rob Nance

DAVE SOUTHALL has been a street performer, acrobat, university lecturer, special effects technician and lately, a TV presenter. He has a degree in electrical engineering, a Ph.D. in electronics, a master's in 3D design, and a shed.

WRITTEN BY MATT RICHARDSON

CYCLE CHASER BIKE PROJECTOR

SHINE FUN ANIMATIONS ON THE STREET THAT MOVE ACCORDING TO YOUR SPEED!

Matt Richardson

MADE PURELY FOR FUN, THE CYCLE CHASER PROJECTS ANIMATIONS ONTO THE GROUND BEHIND YOUR BIKE WHEN YOU RIDE AT NIGHT. The faster you ride, the faster the animation runs. When you slow down, so does the animation. Of course, you can customize this project by changing the animations. You can also dive into the code and change the behavior based on your speed. Perhaps the flames behind you grow longer when you go faster? That's all possible with some clever coding.

Mounted on your bike's rear rack, a battery-operated pico projector gets its video from a Raspberry Pi computer (Figure ). The project code is written with the help of openFrameworks (openframeworks.cc), an open source C++ toolkit for creative coding. I love openFrameworks because it can be used to develop for many different platforms: Windows, Mac, iOS, Android, and of course, Linux. The developers of openFrameworks have done a fantastic job of supporting the Raspberry Pi, so getting up and running is mostly painless ... it just takes a bit of time.

When you use openFrameworks with WiringPi (wiringpi.com), a Raspberry Pi GPIO library, you can read and write to the Pi's pins right from your code. In this project you'll connect a Hall effect sensor, which can sense when there's a magnetic field nearby.

A small magnet attached to your bike wheel triggers the sensor with every revolution, then the code displays the next frame of the animation. This means that if your wheels are turning at 60rpm, your animation will play at 1 frame per second (fps).

After you get it up and running, I encourage you to tinker with the code. How could you take it further? Project your speed in giant numerals on the street, or blinking turn signals? Get hacking, and most importantly, have a safe and fun ride! ◐

Time Required:
3–4 Hours
Cost:
$500–$600

MATERIALS

- » **Raspberry Pi** Maker Shed #MSRPIK2, makershed.com
- » **Pico projector, battery operated** such as the Aaxa P4-X
- » **USB battery pack with micro-USB cable** to power the Raspberry Pi, such as Lenmar #PPW11000UR
- » **HDMI cable, full to mini**
- » **Hookup wire** Maker Shed #MKEE3
- » **Small perf board**
- » **Hall effect sensor** such as Melexis #US5881
- » **Resistor, 10kΩ**
- » **Small magnet**
- » **Electrical tape**
- » **Velcro straps**
- » **Bike with rear rack**

TOOLS

- » **Soldering iron**
- » **Keyboard and monitor** for setup only
- » **3D printer (optional)** for making a mounting "sled"

A

MATT RICHARDSON is a San Francisco-based creative technologist and product evangelist for Raspberry Pi. He's the co-author of *Getting Started with Raspberry Pi* and the author of *Getting Started with BeagleBone*.

Get complete instructions, code, and video, and share your ideas on the project page at makezine.com/go/cycle-chaser.

Skill Builder

TIDBITS & TIPS ON ALMOST ANYTHING

WRITTEN BY JORDAN BUNKER
PHOTOGRAPHED BY HEP SVADJA
ILLUSTRATIONS BY JIM BURKE

Do some skills seem too complex to comprehend? We're here to make things simpler, and to give you a jump-start on learning techniques to push your projects to the next level.

CHOOSING THE RIGHT BATTERY

You've designed the rest of your project, so now it's time to select a battery. There are many types of battery chemistries, and each one has its own different characteristics. So which one to choose? We'll cover the basic features for each battery type in order to help you decide which one is right for you.

For each type below, we've rated the following attributes on a scale of 1–5, with 5 being the best, and 1 being the worst:

SPECIFIC ENERGY: This is the amount of energy in the battery per unit of mass, usually measured in watt-hours per kilogram (W·h/kg).
CYCLE LIFE: This is the number of discharge-charge cycles a battery can be put through before it begins to fail.
SHELF LIFE: This is based on the self-discharge rate of a battery, or how much energy it loses without being used.
PRICE: Battery prices can vary wildly, depending on how much power you want in a given size.

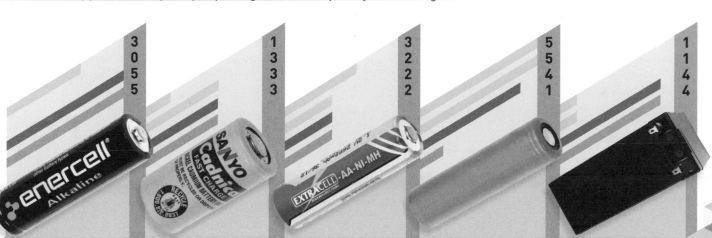

	Dry Cell	NiCad	NiMH	LiFePO₄	Lead-Acid
	3	1	3	5	1
	0	3	2	5	1
	5	3	2	4	4
	5	3	2	1	4

Dry Cell

Dry cell batteries power most small, portable devices. The most common chemistries are zinc-carbon and alkaline. Most are not rechargeable, and are only used for low-power applications. However, their high availability, safety, and long shelf life make them an ideal and versatile choice for small projects.

NiCad

Nickel-cadmium (NiCd or NiCad) batteries excel at maintaining voltage and holding charge when not in use. However, NiCads can fall victim to the dreaded "memory" effect when a partially charged battery is recharged, lowering the future capacity of the battery.

NiMH

Nickel-metal hydride (NiMH) batteries offer a greater power density than NiCad, but they can't handle high discharge rates as well. Though more expensive with a lower cycle life, they are not susceptible to the "memory" effect as NiCads are.

LiFePO₄

Lithium iron phosphate (LiFePO₄) batteries deliver high currents, and have a cycle life in the thousands. LiFePO₄ also won't explode when shorted, as previous lithium ion batteries were known to. Of course, all of that comes at a higher price per battery.

Lead-Acid

A low-cost workhorse, the lead-acid battery is heavy, but it's reliable. Because of their weight, they're often used in non-portable applications such as solar-panel energy storage, vehicle ignition and lights, and backup power.

WORKING WITH SHEET METAL

Thinking of building an enclosure for your latest project? Or maybe you're working on a giant robot to terrorize the neighborhood? Chances are, sheet metal will play a part. Sheet metal comes in all manner of varieties and sizes. Here are some tips and tricks to help you get that shiny sheet into your desired shape.

THICKNESS

One of the most important decisions when working with sheet metal is deciding what thickness you'll need. Similar to wire, sheet metal thickness is measured in gauges, with a higher number indicating a thinner sheet. To measure the thickness, you can use a sheet metal gauge, which will show you thickness in both gauge number and thousandths of an inch. One important note, however, is that ferrous and non-ferrous sheet metals of the same gauge have different thicknesses, so you'll need one gauge for ferrous metals, and one for non-ferrous.

TIP
For small pieces of metal, you can mimic the process below by clamping the metal between wooden blocks in a vise, then hammering them over.

BENDING

Bending sheet metal can be tricky, but with the right tool it's easy. Those who work with it regularly are likely to have a sheet metal bending brake in their workshop, but this tool can be a bit expensive for the hobbyist. Thankfully, there are a couple of wallet-friendly options that can help you get the job done.

Using the edge of your workbench, a length of wood, two clamps, and a mallet, you can fashion a rudimentary bending brake. Mark a bend line and place the sheet metal on the edge of your bench. Next place the wood parallel and slightly behind the bending line. Clamp the wood on top of the metal to the workbench. Finally, bend the sheet up by hand to the angle desired. If you want a sharp 90° bend, tap along the crease with a mallet.

ALUMINUM

COPPER

STEEL

BRASS

GALVANIZED STEEL

CUTTING

Many different tools cut sheet metal, and each has its own strengths and weaknesses. Here are some of the more common tools, but they represent only a small number of options.

SNIPS

Commonly known as "aviation snips" or "tinsnips," these scissor-like tools are great for cutting soft sheet metals such as tin, aluminum, brass, and thin-gauge (24 gauge or thinner) steel. Depending on the cut shape, you'll use left-cut, right-cut, or straight snips, typically indicated by the handle color: red for left, green for right, and yellow for straight. Be sure to insert the metal fully into the throat of the snips for optimal cutting.

BAND SAW

With the appropriate blade, cutting sheet metal on a band saw is fairly straightforward. Cutting metal requires slower blade speeds than cutting wood, but many band saws have multistep pulleys for changing the blade speed.

JIGSAW

A quality jigsaw and the correct metal-cutting blade will make short work of sheet metal cuts. If you need a straight cut, clamp a straight-edge to the sheet to act as a guide for the jigsaw footplate.

NIBBLER

The nibbler is a tool that offers a lot of control over the cut, but at the expense of cut width. Each cut punches out a tiny piece of the sheet metal, and the process is repeated. The nibbler shown here is hand-operated, though drill-powered, electric, and pneumatic versions are also common.

HACKSAW

A hacksaw can cut sheet metal, but its shape limits its turning radius and depth of cut. To prolong blade life, rub wax on the sides above the teeth. For a cleaner cut, put a strip of masking tape on the top and bottom of the sheet to keep chips from scratching the material.

CUTTING ALUMINUM WITH A TABLE SAW

It might sound crazy, but you can cut sheets of aluminum on a table saw. Make sure to use a 60-tooth (or more) carbide-tipped blade, and wax the blade to ensure the cut is well lubricated. Go slowly, proceed with the utmost caution, and wear hearing protection!

DEBURRING

After cutting metal, there's often a sharp edge left. Be sure to remove it! You can buy a fancy deburring tool if you work with sheet metal frequently, but a quick once-over with a file is just as effective. Your fingers will thank you!

INTRO TO SPREAD SPECTRUM R/C

Entering the world of radio-control (R/C) hobbies can be a little daunting. To make it a bit easier to get started, we'll cover the basics of 2.4GHz spread spectrum radio control, and the different pieces of a typical R/C system.

TRANSMITTER

The role of the transmitter is to convert movement of the control sticks into digital signals, which are sent via radio waves to the receiver. Transmitters offer multiple channels to control multiple components. For example, if a transmitter has 6 channels, you can control up to 6 servos or motors. These channels are not to be confused with the sub-frequency "channels" of a radio spectrum.

While older R/C systems communicated over frequencies such as 72MHz and 75MHz, modern systems use the 2.4GHz microwave radio spectrum, plus a collection of clever wireless protocols for more reliability. At the heart of these new R/C systems are Frequency-Hopping Spread Spectrum (FHSS) and Direct-Sequence Spread Spectrum (DSSS) technologies. Both work to nullify the interference and frequency conflicts found in the older R/C systems. While the actual wireless technologies used vary between transmitter brands, all of them include some combination of DSSS and FHSS to avoid interference and frequency conflicts between R/C systems.

OLD R/C

Previous to the 2.4GHz spread spectrum revolution, most R/C systems worked on 27, 50, 53, 72, and 75MHz. With these systems, only one transmitter could be used on a given frequency at a time. Using multiple transmitters on the same frequency would interfere and cause loss of control of the R/C vehicles. Operators also had to be cautious of areas that were prone to noise on the frequency they were using. As a result, almost all R/C systems sold now use the much safer 2.4GHz spread spectrum system.

R/C OVER WI-FI AND BLUETOOTH

Many new multirotors and toy cars can be controlled via Wi-Fi or Bluetooth wireless technologies. One major benefit of this is that you can control the vehicle using apps on smartphones and tablets, eliminating the need to purchase a separate transmitter. The drawback, though, is that range is limited, and you don't get the same tactile feedback from a screen as you do from a typical R/C transmitter's joysticks and switches.

FHSS

Frequency-Hopping Spread Spectrum (FHSS) technology constantly changes the channel that a radio signal is transmitted on, in order to lower the likelihood of signal corruption due to interference on a single channel. The pattern of frequency hopping is pseudo-random, but the transmitter and receiver go through a binding process before any signal is transmitted, ensuring they jump to the same frequency at the same time.

DSSS

Direct-Sequence Spread Spectrum (DSSS) technology spreads the radio signal across a wider range of channels (sub-frequencies) than the old narrowband single-channel systems. This means that even if several of the channels are subject to interference, the signal will still get through on the other channels.

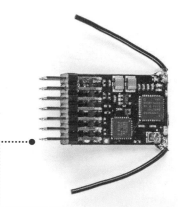

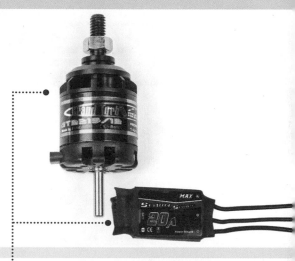

RECEIVER

The receiver collects control-stick data from the transmitter and distributes it to the servos and motors on the R/C vehicle. Transmitters and receivers are frequently sold together, though you can buy them separately. If you do buy them separately, make sure that each unit is compatible, as different brands use different proprietary technologies, and will often not work with one another.

The receiver shown here is a bare board, but most receivers come in plastic cases. The short wires protruding from the receiver are the antennas. If your vehicle is made of metal or carbon fiber, be sure to mount the receiver in a place where the antennas will not be obstructed, as 2.4GHz radio waves will not pass through these materials.

SERVOS

Servos are geared motors designed for precision control over movement. Inside a servo you'll find a circuit board, a small DC motor, and a series of gears. The receiver outputs a pulse-width modulation (PWM) signal to the servo, which the circuit board translates into precise control signals for the DC motor. The circuit board also takes input from a feedback potentiometer attached to the output shaft of the servo to detect its rotation. It then compares the desired shaft position, based on the PWM signal, to the actual position to know which way to turn and when to stop.

ELECTRONIC SPEED CONTROLLERS AND MOTORS

To control a motor, you'll need an electronic speed controller (ESC). Its job is to take the low-power signal from a receiver and turn it into high-current control signals for driving a motor. ESCs come in two varieties depending on your motor type: brushless and brushed. Brushed speed controllers send a PWM signal to a brushed motor, while brushless speed controllers switch power between the three leads required for brushless motors.

When selecting an ESC, you'll need to know the maximum current rating for your motor. The maximum current rating on your ESC should be 5 or 10 amps higher than the maximum rated current of your motor, as motors often draw more current than they're rated for. Many ESCs are also programmable, allowing you to set different attributes for controlling your motor.

KNOW YOUR RIVETS

Rivets are one of the oldest, most reliable methods of permanently fastening two materials together. The principle is simple: a shaft with a head on one end is inserted into a hole, and the opposite end is deformed to hold the materials together. Like many simple inventions, time has brought new and more complex features to rivet technology, which has resulted in a variety of different types of rivets. Here we'll cover two of the most common rivets used in the home workshop.

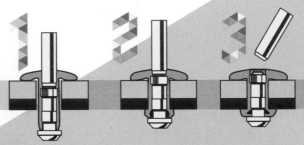

EASY RIVETING

Have hundreds of blind rivets to install? Consider buying a pneumatic rivet gun, or a rivet gun attachment for your drill. You'll work much faster and avoid the strain and pain from squeezing a hand riveter all day.

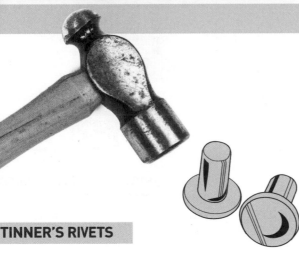

BLIND RIVETS

Blind rivets are the fastest and easiest rivets to use in a project. Often called "pop" rivets (after a popular rivet brand), they can be installed completely from one side, unlike solid rivets or bolts that require access to both sides of the material. If set correctly, blind rivets produce reliable fastenings without all the hammering involved in solid rivet installation.

Blind rivets are made of two preassembled pieces: the tubular rivet body and the setting mandrel. Both pieces are inserted into a hole, and a special tool (appropriately called a "riveter") is placed over the mandrel. Squeezing the grips together causes the mandrel to be pulled into the rivet body, which expands on the other side of the material. If the mandrel breaks off of the completed rivet, the rivet has been installed correctly.

TINNER'S RIVETS

Used for thin sheet metal, *tinner's rivets* are solid rivets, usually made of soft iron or steel and have a flat head. To install a tinner's rivet, the rivet is inserted from the underside through the materials to be fastened, and the flat head is placed on top of a large flat piece of metal (like an anvil). A tool called a rivet set, which has a dished depression slightly larger than the rivet diameter in its end, is placed on the shaft of the rivet and struck by a hammer repeatedly to peen the rivet shaft into a flattened, rounded head.

RIVET MATERIAL

The type of metal used in the rivet you choose is important, and should be matched to the sheet metal being joined whenever possible. If leather is being fastened with a rivet, make sure that the rivet is made of aluminum, copper, or brass, as the moisture in leather can cause steel rivets to rust.

MAKESHIFT RIVETS

If you don't have a rivet set or tinner's rivets, you can use a roofing nail and a ball peen hammer to create a makeshift rivet. Place the nail in the hole and cut the nail down to 1.5 times its diameter. Place the material with the nail head facedown on your striking surface. Using the flat face of a ball peen hammer, give the end of the nail several strikes to set the rivet and begin deforming the metal, then use the ball side of the hammer to work your way around the rivet in a circular pattern, shaping the rivet into a head.

SPRAY PAINT TECHNIQUES

Spray paint is a convenient way to add a touch of color to your project. It's available in just about any color you'd want, and can be purchased at your local big-box hardware store. Using spray paint may seem like a straightforward process, but if you're not familiar with it, the results can be disappointing. Here are some tips and tricks to ensure that your paint job comes out as beautiful as you'd expect.

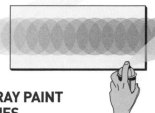

BASIC SPRAY PAINT TECHNIQUES

First, make sure that the surface to be painted is clean, free of rust and debris, and smooth. Any surface bumps or imperfections will show through the paint, so use sandpaper or steel wool to clean and smooth the surface, then use a lint-free cloth to remove any remaining dust.

Start with a can of spray paint that is room temperature and shake it for 3 or 4 minutes to mix the paint thoroughly. You can't shake the can too much, but you can shake it too little! Be sure to shake the can occasionally while painting as well.

Press down on the tip with the can about 10" to 12" away from the surface, aimed to one side, then, in one fluid motion, sweep the paint horizontally across the surface and release the tip when you reach the other side. Use your entire arm to move the can, not just your wrist, and be sure to start the spray before reaching the surface, and release after passing it.

For a large surface where multiple passes are required, overlap your passes slightly. Several light coats, allowing the paint to dry in between, will look much better than one heavy coat. Different paints dry at different rates, so read the instructions on the back of the can for drying times. Be patient, apply multiple light, smooth layers of paint (usually at least three), and your paint job will come out looking professional.

When you're done painting, turn the can upside-down and press the tip until the spray becomes clear. This will clear the tip and prevent paint from drying inside it.

PAINTING SMALL OBJECTS

If you're painting something small, use a paint booth to limit overspray. You can set up a makeshift paint booth using a cardboard box on its side. Setting the object on a turntable will allow you to spin it without touching it so that you can paint from all angles.

USING A SPRAY GUN

If you have a large painting job, and have a compressor, consider buying a spray gun. Instead of using compressed gas in a can, a spray gun uses the air delivered from your compressor to suck paint from a paint basin and blow it through a nozzle, creating a fine spray.

USE A RESPIRATOR!

Spray paint fumes are toxic, and cheap dust masks offer very little protection. Be sure to spray-paint in a well-ventilated area where paint fumes won't build up. A respirator with replaceable filters costs just $20 or $30, and will survive many paint projects. You'll save money in the long run, and it's much cheaper than going to the doctor for respiratory problems.

REVIVING OLD CANS OF PAINT

Occasionally, a used can of spray paint will refuse to spray. Remove the nozzle and soak it overnight in a solvent such as mineral spirits or paint thinner to dissolve the paint. If that doesn't work, try putting the can in a bucket of hot (not boiling!) water. This will reduce the viscosity of the paint, and help increase the pressure in the can.

RETRO R/C RACER

Now build your own vintage-styled R/C car! This project from *Make:* Volume 11 walks you through the steps to build an R/C vehicle from scratch, with a sheet-metal body based on 1930s midget racers. You'll use all the tools and materials covered in this section (with a few changes to keep pace with technology) to create your racer, but they can also be applied to all manner of similar projects.
makezine.com/go/retro-rc-racer

INSTALL A BATTERY

CONNECT THE R/C PARTS

RIVET THE SHEETS TOGETHER

BEND THE ALUMINUM

SPRAY-PAINT THE BODY

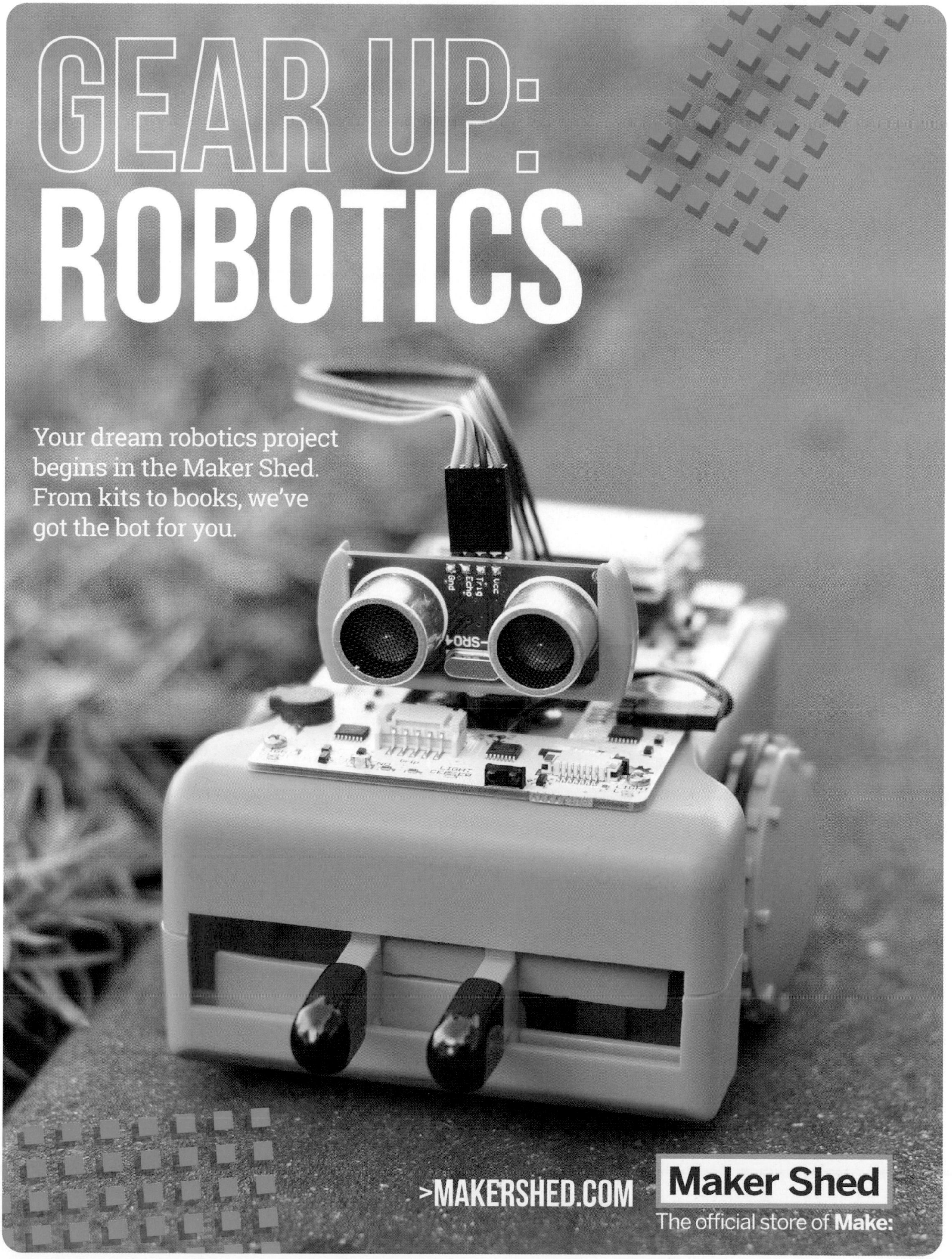
GEAR UP:
ROBOTICS
Your dream robotics project begins in the Maker Shed. From kits to books, we've got the bot for you.
>MAKERSHED.COM
Maker Shed
The official store of Make:

Building Your Own
BB-8

LOOK OUT, R2! Everyone's favorite astromech is facing stiff competition from a new addition to the *Star Wars* universe named BB-8. Aside from being impossibly cute, BB-8 is a puzzle of engineering that features an R2-style head gliding on a rolling, spherical body. As a fan, it's impossible not to want one. As a Maker, it's impossible not to try to craft one. Want to get started? Here are 3 different approaches for building your own.

Christian Poulsen's BB-8

Pocket-Sized Roller:

Time Required:
1 Day
Cost:
$150–$200

Materials
» Sphero 1.0 robot toy
» Plasti Dip spray, GunMetal Gray
» Spray paint: orange, white, and clear enamel
» Neodymium ring magnet, ¾"
» Neodymium disc magnet, ⅜"
» Felt pad, .090" × ⅜" diameter, adhesive-backed
» Polyurethane foam
» Wood spackle
» Masking tape
» Sharpie marker

Tools
» Hacksaw
» Bench vise
» 3D modeling software
» Glue gun
» Hobby knife
» CNC mill (optional)
» Laser cutter (optional)
» Drill press (optional)

POCKET-SIZED ROLLER
Written by Christian Poulsen

PROJECT #1

Hack a Sphero to replicate the galaxy's adorable new droid

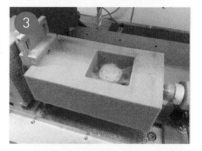

CHRISTIAN POULSEN is a recent graduate from Brigham Young University, with a degree in industrial design. In addition to his freelance design work, he enjoys restoring his '70s Porsche named Sebastian.

AS AN INDUSTRIAL DESIGNER, I'VE BEEN PARTICULARLY FASCINATED BY PRODUCTS THAT HAVE PERSONALITY AND EMOTION. I think the droids in *Star Wars* have always been really great at capturing a character without facial expression and drawing you in. And BB-8 was no different when they brought it out onstage for the first time at *Star Wars* Celebration.

As I watched it roll around, my only thought was, "Need!" So I made one.

Part of what I really enjoyed about the process of making it was the timeline. Most projects I work on end up taking weeks, if not months, to finish. With BB-8, I pushed myself to make it in a day. As a result, the surfacing and paint is less than perfect, which I'm serendipitously calling "weathered." But I was able to stick to my goal, and made the whole thing in a matter of hours. I wanted to capture the character and personality of the real robot as simply as possible, and I'm really happy with how it turned out.

1. SPLIT SPHERO 1.0
Divide the Sphero toy along the seam using a hacksaw, being careful not to saw through the internal chassis. (I've never split open the Sphero 2.0, so I'm not sure if it will work on that model.)

2. EMBED THE MAGNET
The internal chassis of the Sphero has a post that pushes against the top of the sphere and acts as somewhat of a shock absorber — attach the ¾" neodymium magnet ring to this post. Clean the edges of the split sphere with a hobby knife and, with the newly magnetized structure inside, reattach the 2 halves with hot glue.

3. DESIGN AND FABRICATE THE HEAD
Create in your preferred 3D modeling program or simply carve by hand. I built it in Rhino over a

screenshot from the trailer to make sure I had the proportions right. I milled out my design on a CNC using polyurethane foam, then used wood spackle to finish the surface.

4. MASK AND PAINT
Cut the designs out of tape to mask off the different colors. I went a little overboard and laser-cut mine, and used gray Plasti Dip for the base coat because it's more opaque than regular spray paint and I wanted to block out the Sphero's flashing LEDs (a glowing green droid just wouldn't work for me). After the base coat is dry, cover with orange spray paint.

5. APPLY FINISHING COATS
Mask out the orange and add a coat of white paint. Add the details with a Sharpie and finish it with a final coat of clear enamel. The topcoat is particularly important to get just right, to reduce friction as much as possible between the head and body.

6. ADD MAGNETS AND FELT
Once the paint is completely dry, embed a ⅜" neodymium disc magnet in the head. Add a small adhesive felt pad so that it can slide easily across the surface of the body. I used a drill press to sink the magnet in just deep enough to be flush with the surface and glued it in place before attaching the ⅜"-diameter felt pad over it.

The droid is controlled by Bluetooth through the Sphero app, but also has a mind of its own. Because it's now more top-heavy and tends to lean, the gyroscope will try to correct the lean, and it will keep on rolling in whatever direction it's pushed. Maybe adding weight in the bottom will tame it, or maybe BB-8 needs a restraining bolt!

For a more detailed guide to making this project, visit makezine.com/go/sphero-bb8.

Written by Kurt Zimmerman

POWER WHEELS AND HARD DRIVES

PROJECT #2

An experienced R2-D2 builder shares the ups and downs of his BB-8 project

Kurt began construction on BB-8's head and shell as far back as November of 2014, when the character made its first on-screen appearance.

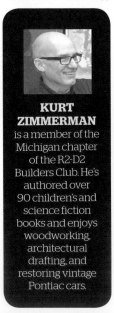

KURT ZIMMERMAN is a member of the Michigan chapter of the R2-D2 Builders Club. He's authored over 90 children's and science fiction books and enjoys woodworking, architectural drafting, and restoring vintage Pontiac cars.

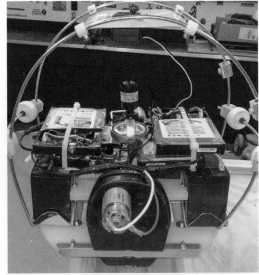

Inside is a mix of repurposed materials, including hard drives, cutting boards, brush handles, and two Power Wheels motors.

WHAT THE HECK WAS THAT?

This was my first impression of BB-8 when a 3-second glimpse of it was revealed during the first "teaser trailer" for *Star Wars* Episode 7. No one had seen a droid such as this. A dome bouncing along on a rotating ball.

I had to have one.

I reasoned that I could drive a ball with a remote control tank inside. The trick would be to keep the dome from falling off. Three weeks later, I had an outer shell with a finished Styrofoam dome. The dome had five cannibalized computer mice underneath, their rubber balls replaced with ⅞" steel bearings. The inner tank had a framework attached, supporting an array of five magnet columns, spring-loaded to keep them in contact with the inner ball while the tank crawled around the interior. This design failed for two reasons. Firstly, the tank only touched the ball on its four corners, so there was little traction. Secondly, the inertia of the droid when it stopped kept the lower ball rolling, causing the upper dome to tip off. I needed to increase the traction and come up with a way to stabilize the ball.

The solution I came up with was a custom drive system that puts the weight and traction directly in the bottom of the lower ball. Two rubber drive wheels, powered by two Power Wheels motors and gearboxes, and controlled by radio with a Sabertooth 2×25 motor controller, would give me the control and traction I needed. I would be able to turn the drive system 360° and have enough power to overcome the drag of the magnets at the top of the ball. I added a sliding platform so that I could move the magnet array forward and backward inside the lower ball. A 360° servo rotates the magnet platform and the upper dome.

For the enclosure I used 18" polycarbonate globes from edee.com. (The community now feels 20" is more accurate.) Because the globes come with a hole already cut out, and gaps occur when you cut out the outer shell pieces, three globes are needed.

To stabilize the droid, I purchased the only "old school" gyroscope I could find, but it didn't have enough rotating mass to stabilize such a large ball. To add more rotating mass, I built four additional gyroscopes out of discarded computer hard drives, stacking four discs in each drive and spinning them with a control circuit, powered by a battery.

I also tried to replicate the different emblems on the sides of the body. And now that we have better pictures of BB-8, I plan to rework the outer shell to make it even more movie-accurate.

To see the full, finished build of Zimmerman's BB-8, visit makezine.com/go/zimmerman-bb8.

Kurt Zimmerman

WHEN THE *STAR WARS: THE FORCE AWAKENS* TRAILER WAS RELEASED, THE INTERNET WAS BUZZING ABOUT THIS NEW ROUND DROID.
With a renewed interest in practical effects, our curiosity was piqued. "Are they going to actually make one?" This question was answered at the *Star Wars* Celebration. Our curiosity became an obsession as we watched BB-8 roll around and size up R2-D2. The ball droid was real, and hackers and Makers around the world thought they had the answer as to how it worked. Concept art, 3D prints, and prototypes started popping up on YouTube and we knew we had to throw our hat into the ring.

Since there are others trying to crack the BB-8 we wanted to work quickly on our proposed solution. With the help of ServoCity and their online CAD models, we quickly leveraged the Actobotics line to build the inner structure. For some of the detail pieces and internal components, we turned to 3D printing to give us quick results. The Loveland CreatorSpace provided us a base camp and excellent resources to complete this build.

The inspiration for our mechanical build was the Sphero. We had a hunch when the first videos were released that Sphero was involved somehow. We popped one open and began to replicate the mechanism. To add the head and motion we took cues from an older Sphero model. The original Sphero has a singular upper support, whereas Sphero 2.0 has two passive upper wheels to maintain support. Our design combines aspects of both, using upper wheels for stability and a central support to attach the head via magnets.

The shell is 2 custom-molded ⅜"-thick polycarbonate hemispheres. We decided to go this route to ensure roundness, strength, and enough thickness to mount LEDs in the moving shell.

The team behind our build consists of four Makers with a shared passion for robotics and *Star Wars* alike, and their obligatory dog companion. I was responsible for the mechanism that brings the BB-8 to life. Pamela Cortez, another SparkFun engineer, developed the brains of the droid, which consist of Raspberry Pi and Sphero. She also focused on giving our BB-8 a voice, which includes some Easter egg sounds from your other favorite droids! The aesthetics of the robot, including the floating head and movie-worthy details, were applied by Allison Cavis and Maurice Woods, two artists-turned-engineers whose sculpting and casting experience make the droid's appearance unmistakable.

This BB-8 design is fully documented and open source! For the complete build, visit makezine.com/go/open-bb8.

Written by Casey Kuhns

PI-POWERED SUPER-SPHERO

PROJECT #3

A team of engineers and artists methodically creates a BB-8 that looks, moves, and sounds like the real thing

The team (from the left) includes Pamela Cortez, Allison Cavis, Casey Kuhns, Maurice Woods, and Felipe.

An inner structure was first carefully designed in software, based on the construction of a Sphero toy.

CASEY KUHNS
is a SparkFun engineer by day, and a gearhead by night. From Boulder Colorado, he has built space payloads, restored a jet engine, and fought combat robots.

Chukudu Wooden Scooter

Build the burly Congolese workhorse that can haul hundreds of pounds of cargo

Written by Doug Bradbury

Time Required:
A Weekend
Cost:
$100–$150

DOUG BRADBURY
is director of software services for 8th Light in Chicago. His interest in social justice has brought him to Africa on several occasions. He met our chukudu designer AIMÉ NSHIMIYIMANA (pictured here with *Make:* Volume 40) in the Kiziba refugee camp in Rwanda.

Ryan Goebel

Materials

» **Dimensional lumber: 2×10, 10' length (1) and 2×4, 4' lengths (2)**
» **Hardwood dowel, 2" diameter, 48" length**
» **Steel wheels, 10"×2.75", with ⁵⁄₈" diameter hub (2)**
» **Threaded rod, ⁵⁄₈", 8" lengths (2)**
» **Lock washers, nylon, ⁵⁄₈" (4)**
» **Nails, 2"**
» **Staples, ½" wide, 1" long**
» **Rubber tie-down straps, adjustable, 36" (4)**
» **Bicycle inner tube, salvaged**
» **Bicycle tire, salvaged**
» **Wood glue**

TOOLS

» **Jigsaw**
» **Handsaw**
» **Hammer**
» **Chisel, 1"**
» **Hole saw, 2"**
» **Drill and bits**
» **Hacksaw or reciprocating saw with metal blade**
» **Wrench**
» **Router (optional) with ¾" straight and roundover bits**
» **Sandpaper**
» **Clamps or weights**

FOR THE LAST 20 YEARS, TWO CIVIL WARS AND PROLONGED VIOLENCE have left the Democratic Republic of the Congo unsafe for many. Since 1994, it's estimated that 3 million people have been killed there.

In western Rwanda, just 60 kilometers away, sits the Kiziba refugee camp, where 18,000 Congolese wait hoping for peace so they can return to their homeland. Food and firewood rations are provided monthly by the United Nations, and the responsibility for moving these loads typically falls to young boys equipped with a handcrafted, all-wood scooter called a *chukudu* (choo-KOO-doo).

Often called the pickup truck of the Congo, chukudus can haul loads upward of 500 pounds. They're pushed up hills and ridden down the other side. Today, the chukudu makers in Kiziba carve these scooters from eucalyptus trees with nothing but a machete.

After visiting the camp, I wanted to honor the ingenuity and resourcefulness of the Congolese refugees by building my own version of the chukudu. You can build it with common hardware.

ABOUT THE DESIGNER

Aimé Nshimiyimana grew up in the Kiziba refugee camp, where he studied auto mechanics and drafting. Now 24, he walks 4 hours round trip to town each day to make a few dollars fixing cars. He wants badly to study at a university. When I asked him about the chukudu, he enthusiastically drew plans and explained how they're built. This project is an adaptation of his design (Figure Ⓐ).

To help Aimé and others like him complete their education, you can support International Teams at give.iteams.us/give. Specify "Impact Rwanda - Kiziba Education" with your donation.

1. PREPARE THE DECK

Cut the 2×10 to 5' long and round the nose with a jigsaw, using a 10" wheel as your template. Also mark the board through the center of the wheel, then cut a 2" hole on your mark (Figure Ⓑ).

Cut an 8"×3" notch centered in the rear end of the deck. Optionally, use a router to cut a ¾" groove ¼" deep across the bottom, 2" from the rear end. This helps keep the rear axle in place. (Without a router, nailing and stapling will do.)

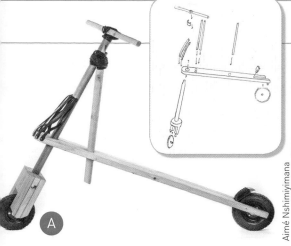

A

B

Aimé Nshimiyimana

C

Andy Waters and Doug Bradbury

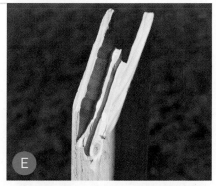

D

2. BUILD THE FORK

Cut three 12" lengths of 2×10. Cut 2 of these at 45° on one end, for the outside pieces. Cut matching 6"×3" notches out of all 3 boards (Figure **C**). Then cut a ¾" groove ½" deep across the end of the middle piece, to accept the axle.

Square off one end of the 2" dowel: Cut an ⅛" groove on each side 3½" from the end, then use a chisel to split off all 4 sides, down to your grooves.

On the middle 2×10 cut a notch 3½" deep and 1¾" wide on the other end, to accept the squared end of your steering shaft as tightly as possible.

At the other end of the shaft, square off 2 sides (Figure **D**) to accommodate the handlebars. Be sure your flats line up on both ends of the shaft.

Glue and clamp the three 2×10s together, sandwiching the steering shaft tightly in between.

3. MAKE THE STEERING SHAFT SUPPORT

Exactly 9" behind the center of the steering shaft hole, cut a 3½"×1⅞" hole in the deck, at a 20° angle, to accommodate a 4'-long 2×4 so it meets the steering shaft 6"–8" from the top. A tight fit is important, so start small and widen as needed.

Now shape the support: Cut a 70° miter at the top of the 2×4, then shape a concave in the front face using a router or chisels (Figure **E**).

Install the support into the deck hole. It can stick through the bottom or be trimmed flush.

4. SHAPE THE HANDLEBARS

Mark a 30" 2×4 at 12" from either end. Then mark it lengthwise 1" from either side. Use a jigsaw or handsaw to cut away the 4 corner portions, leaving a 6" hub in the middle and 2 handles.

Find the center of the hub and place it atop the steering shaft. Trace the outline of the shaft and cut out that hole (inside the lines!) with a drill and jigsaw (Figure **F**). The fit should be tight.

Use sandpaper or a router with a roundover bit to round off the sharp edges of the handle. Don't mount the handlebars yet!

5. MAKE THE BRAKE

Your brake will be made just like the ones in Kiziba. Cut a 7" section of the bike tire. Staple or nail a scrap of wood inside one end (Figure **G**).

Nail the other end inside the notch you cut in the deck for the rear wheel. The curve of the bike tire will cause the brake to sit up off the wheel. The rider can then step back and push the brake into the wheel to slow the chukudu.

6. MOUNT THE WHEELS

Cut two 8" lengths of ⅝" threaded rod for axles. Mount each wheel in the center of an axle and secure it with 2 nylon lock nuts. Don't overtighten.

Secure the axles to the deck and the fork with long staples or 2" nails bent over the axle. Drive a nail or staple at the end of each axle to prevent it from moving off-center (Figure **H**).

7. MOUNT THE SUSPENSION AND BARS

The Congolese scavenge rubber from old road tires. We'll use 4 rubber tie-down straps.

Remove the hooks, and nail the ends of all 4 straps around the nose of the deck. Thread the steering shaft up through its hole in the deck. Pull up the center of each strap and nail them to the steering shaft (Figure **I**). Set the height so the deck hangs 6"–8" above the fork, so that when the chukudu is loaded it won't bottom out.

Use the reclaimed bike tire to wrap and hold the steering shaft to the support. Nail or staple the tire into the support on either side. Finally, mount the handlebars (Figure **J**). Use an old inner tube to tie them on, and/or a shim to tighten the joint.

RIDING THE CHUKUDU

Customize your deck for hauling different types of cargo, or as a knee scooter. A pad made from an old flip-flop lets you place a knee on the deck.

Because of its size and mass, the chukudu can be difficult to handle. Be sure to wear a helmet — and have an escape plan if you need to bail! ⊘

E

F

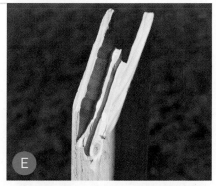

G

H

I

J

Get more photos and build tips at makezine.com/go/diy-chukudu-scooter.

Million Color HSL
Flashlight

Written by Dan Rasmussen

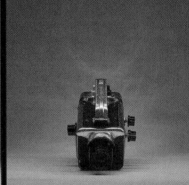

Hep Svadja

Bring fun back to the flashlight with Arduino and full-color NeoPixel LEDs

DAN RASMUSSEN
is an avid collector, fixer, and hacker of vintage technology. He's a software engineer who lives in Groton, Massachusetts with his wife and three kids.

Time Required:
4–6 Hours
Cost:
$40–$60

Materials

- **A big 6V lantern flashlight to hack** old or new
- **Arduino Pro Mini 328 microcontroller board, 5V, 16MHz** Maker Shed #MKSF8, makershed.com, or Adafruit #2378, adafruit.com
- **Right angle male headers, 6 pin** cut them from Adafruit #1540 or similar.
- **NeoPixel RGB LED ring,** Adafruit #1643
- **Potentiometers, 10kΩ (3)**
- **Hookup wire, 22 gauge solid core** Maker Shed #MKEE3 or Adafruit #1311. It's nice to have lots of colors for this project.
- **Batteries, NiMH rechargeable, AA size (4) and charger** Do not use alkaline batteries.
- **Battery holder, 4xAA**
- **Resistor, 300Ω, ¼W**
- **Capacitor, 1,000µF, 6.3V or higher**
- **Rotary switch, 10-position** such as SparkFun #13253
- **SparkFun Rotary Switch Breakout board (optional)** SparkFun #13098
- **Potentiometer knobs (4)** such as Adafruit #2046
- **Heat-shrink tubing** various diameters
- **Tic Tacs, any flavor you like**
- **Solder**

Tools

- **Soldering iron, temperature controlled** such as Maker Shed #MKME01, set to 700°F
- **Wire strippers** such as Adafruit #527 or, for fun, Maker Shed #MKLTM2-ES4
- **FTDI Serial TTL-232 USB cable** such as Adafruit #70
- **Dental pick (optional)** handy for attaching wires to posts
- **Hot glue gun**
- **Computer with Arduino IDE software** free download from arduino.cc/downloads
- **Project code** Download the Arduino sketch *HSLFlashlight. ino* from the project page, makezine.com/go/million-color-hsl-flashlight.
- **Hobby knife**
- **Adhesive tape**
- **Heat gun or butane lighter** for heat-shrink tubing
- **Drill and drill bits**
- **Digital voltmeter or multimeter**
- **Sharpie marker**

WHEN I WAS A KID THERE WERE NO SMART-PHONES, INTERNET, HOME COMPUTERS, OR ARDUINOS. Sure, we had TV and radio, but — believe it or not — we had fun with flashlights too. They often came with colored lenses, or we made our own with paper or plastic.

These days flashlights are mostly for practical purposes and they usually only make white light. Even high-end flashlights are pretty much the same thing, just super bright and rugged for inspecting the levee during the hurricane. Bright and rugged is great, but a one-button, small, efficient, vanilla light isn't much fun, is it? Let's bring the fun back to flashlights.

THE HSL FLASHLIGHT

In this project you'll hack an old-style 6-volt lantern to become a Million Color HSL Flashlight. It's big and bulky and different. Not quite as bright as some, but it has crazy color that's so easy to set — just dial your way around the rainbow. Lots of knobs and modes, and you can write code for it too.

The great thing about HSL is that it's an intuitive way to select a color. HSL stands for *hue*, *saturation*, and *lightness*. It's like the 360-degree color pickers available in many computer applications. Our HSL flashlight has a knob for each component: hue selects the color by allowing you to dial your way around the rainbow, saturation selects how deep and rich the colors are (fully desaturated is white, fully saturated is pure color), and lightness behaves like a dimmer.

The flashlight also has a 10-position mode switch, just for fun. The following are the modes

I programmed — watch the video on the project page at makezine.com/go/million-color-hsl-flashlight for a demonstration — but you can always program your own!

1. White
2. Manual HSL selection
3. Auto hue: rotates through the color spectrum (hue knob controls speed)
4. Multicolor all-pixel auto-rotate
5. Multicolor tri-pixel auto-rotate
6. Cylon (aka Larson Scanner)
7. Full-color strobe
8. Alternate pixel multicolor 180° color alternation
9. Half-moon multicolor 180° color alternation
10. Alternate pixel-pair multicolor 180° color alternation

HARDWARE

Arduino is a great platform for prototyping your projects. Once you're ready to build, though, the standard Arduino board is kind of big and its jumper wires provide only fragile connections. So I used the Arduino Pro Mini: inexpensive, very small, and reliable.

I chose Adafruit NeoPixel RGB LEDs because their 12-element ring fits nicely in the reflector bowl in most 6V lanterns, and they come with an Arduino library that's easy to use.

Old-school "6-volt lantern" flashlights are perfect for this project because the reflector bowl and the battery compartment are both huge. Cool old lanterns are easy to find at flea markets or eBay, but you can also buy new ones.

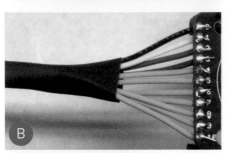

1. SOLDER HEADERS ON THE ARDUINO

Tack a single pin of the 6-pin header, then check to make sure it's flush. If not, reheat and adjust.

Now solder the other pins. Rework the tacked pin and add solder if necessary.

2. PROGRAM THE ARDUINO

Download the Arduino sketch *HSLFlashlight. ino* from the project page, makezine.com/go/million-color-hsl-flashlight. Connect the Arduino Pro Mini to your computer using your FTDI cable or other compatible device. Open the Arduino IDE software and select the correct board from the Tools→Boards menu. Then open the sketch and verify/compile/upload it to the Arduino.

This tiny board has an LED attached to pin 13. The HSL program will flash it 5 times when it starts. This is a good way to verify that the program has been properly uploaded. (Note that the upload process itself will flash this LED a couple of times.)

Disconnect the Arduino and set it aside.

3. WIRE THE 10-POSITION ROTARY SWITCH

The optional breakout board makes it easier to attach and manage the switch's 11 wires. Tack one pin, make sure it's flush, then solder all the connections (Figure A).

Cut eleven 8" lengths of 22-gauge wire. Use different colors of wire, or use a light color and mark the wires with a Sharpie to indicate the pin number. Strip about ⅛" of insulation from each end.

I always use black for common/ground and red for raw power. Here I attached black to the common pin, then used different colors for the first few positions (Figure B). After that I used all gray, and marked the other end of the gray wires with dots (4 dots for position 4, 5 for position 5, and so on).

Tame the mess of wires with some heat-shrink tubing. I covered about 60% of the wires.

If you omit the breakout board, solder the wires directly to the 10-position pot.

4. WIRE THE POTENTIOMETERS

Cut three 8" lengths of wire for ground (black), 3 more for Vcc (I used orange), and 3 for analog input (blue). Connect one set of wires to each potentiometer as shown (Figure C). First wrap the wire around the post, then solder it .

Mark the potentiometers as 1, 2, and 3, then mark the other end of each analog input wire with 1, 2, or 3 dots.

5. PREPARE THE NEOPIXEL RING

Cut about 12" each of black, red, and white wire and solder these to the NeoPixel's Power, Ground, and Signal In pads, routing the wires from the top as shown (Figure D).

> **CAUTION:** Some flashlights (especially old ones) use metal reflector bowls, so it's important to insulate these connections. I simply covered them with small blobs of hot glue.

Hot-glue the NeoPixel to the bowl of your flashlight and route the wires through the old bulb hole.

Solder the 300Ω resistor to the other end of your signal wire (white), and insulate with heat-shrink, leaving about ⅛" exposed.

> **IMPORTANT:** Before connecting a NeoPixel strip to power, connect a large capacitor (1,000µF, 6.3V or higher) across the + and – terminals. This prevents the initial onrush of current from damaging the pixels.

6. WIRE THE ARDUINO

Solder a 12" red wire to the Arduino's RAW power input pin; a 12" black wire to the adjacent ground (GND) pin; and a 12" orange wire to the regulated power (VCC) pin.

Solder the 3 blue wires from the potentiometers to the Arduino's analog inputs A1, A2, and A3 (indicated by the red box in Figure E), matching the dots you marked. A1 is for hue, A2 for lightness, and A3 for saturation.

Dan Rasmussen

Solder the 10 signal wires from the switch to the Arduino's digital inputs 2 through 11, matching position 2 on the switch to Arduino pin 2, position 3 to pin 3, etc. Match position 1 to pin 11.

Solder the 300Ω resistor on the NeoPixel signal wire to digital input 12 (Figure **F**).

7. MANAGE POWER, GROUND, AND VCC WIRES

Tie all 3 raw power wires (red) together as shown — from the Arduino, the NeoPixel, and the battery pack — then solder and finish with heat-shrink (Figure **G**).

Connect all the VCC power wires (orange) together the same way. Then all the ground wires (black).

8. INTEGRATE FLASHLIGHT'S ON/OFF SWITCH

Find the wires that go to each side of the switch in your flashlight. Cut them and strip ½" from the ends.

Cut the red wire from your battery pack and strip ½" from each end. Solder each end to one side of the flashlight switch. Patch in more red wire as necessary to give you some flexibility (Figure **H**).

9. MOUNT THE ARDUINO IN THE TIC TAC BOX

Empty the Tic Tac box, remove the white plastic dispenser, and cut it down a bit to accommodate the bundle of wires. Then insert the Arduino into the box and seal it with strong tape (Figure **I**). This will insulate the Arduino from any metal inside the flashlight.

10. TEST

No doubt you're anxious to see if this thing is going to work! First, center all 3 pots (to be sure the lightness is not all the way down.) Load 4 fully charged NiMH batteries, and power up the electronics by turning on the flashlight switch. Watch for the 5 flashes on the Arduino's LED. Shortly after that you should see some activity from the NeoPixel ring — no matter where the mode switch is set.

If you see no activity on the Arduino, then there's probably a short somewhere — turn off the power right away and find the short. In my case it was the NeoPixel ring shorting against the metal bowl of my flashlight.

11. MOUNT THE ELECTRONICS IN THE FLASHLIGHT

Drill 3 holes in the flashlight case for the pots and one for the rotary switch, then mount them and attach the knobs. I put the HSL controls on one side and the rotary mode switch on the other (Figure **J**).

Now stuff the Arduino, batteries, and wires into the flashlight. These 6V lanterns have plenty of room. You might find that all your heat-shrink is making the wires hard to bend. Just go slowly and it will all work (believe me, it's better than a rat's nest of wires).

CONGRATULATIONS

You've built a fully programmable Million Color HSL Flashlight that will keep you (and maybe even your kids) entertained for years. We'd love to see how you modify it, and what new modes you come up with! ◗

3D-Print a Badass R/C Race Car

Written by Taylor Alexander

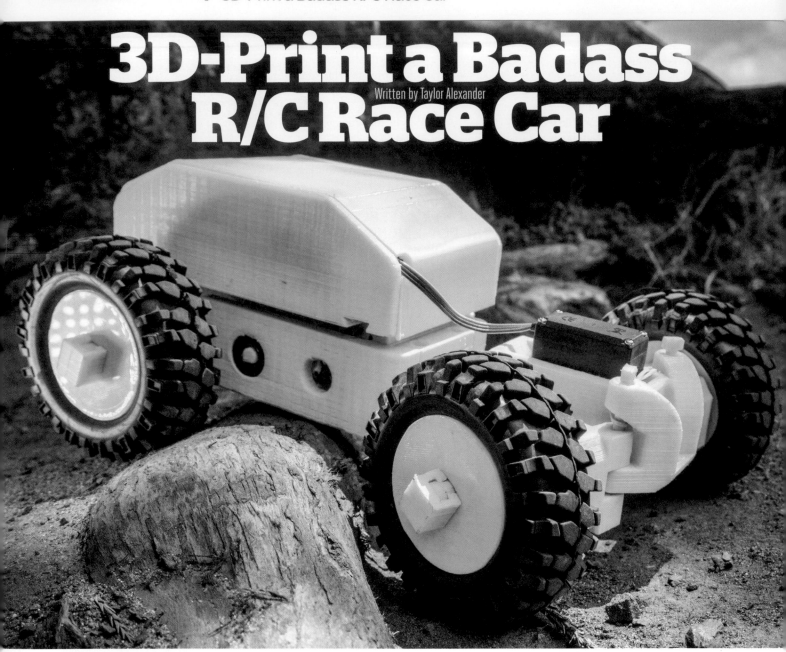

Make this little ripper with a brushless motor, Flutter wireless microcontrollers, and a few bucks of filament

TAYLOR ALEXANDER is a lifelong tinkerer who has been interested in robotics since he was 11 years old. He now runs Flutter Wireless, an electronics company focused on hobbyist electronics and robotics. flutterwireless.com

3D PRINTERS, I THINK, ARE A BIGGER DEAL THAN MOST PEOPLE REALIZE. For the first time in human history, we have a way to produce high-quality, complex parts at nearly zero marginal cost. Case in point: This 3D-printed car only requires $10 in plastic material, 24 hours of machine time, and $1 in electricity. And if the machine could be fed with industrial plastic pellets instead of custom-made filament, it would only be $1 in plastic.

A CNC machine has many automatic features and makes very high-quality parts, but it still takes a lot of skilled labor to use. 3D printers, on the other hand, take no labor at all to produce complex parts. That's a very new thing, and has the potential to do for mechanical goods what the printing press did for books — eliminate essentially all human labor required to make stuff, and drastically lower the cost of living. What if we could print robots that could do all the hard work in society?

I wanted to inspire people to use 3D printers to build real robots, so I built this car — I call it the Scout — to show off some of the 3D-printed

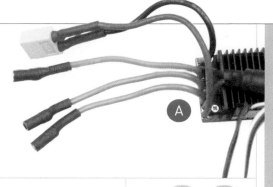

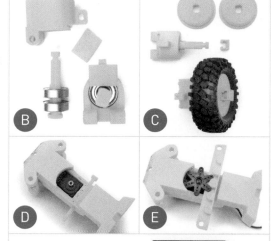

robotics concepts I've been developing. First, the car isn't a weakling. It can take repeated jumps and crashes with no damage. If it does break, it's easy to repair in minutes, thanks to its snap-together design. And the design is open source, so it can mature over time and inspire new variants. It uses a minimum of non-printed parts — only the tires, bearings, motors, batteries, and electronics aren't printed — and it doesn't even use screws!

The wireless radio in the car and remote is my own design — a board called Flutter. Flutter runs Arduino code and has a 1-kilometer range. Since it runs Arduino, the car can be modified with lights, sensors, speakers, and more. I hope to develop motor control and battery management boards for robots as well, so that one coherent open source system can be used to build almost any robot.

Assembling the Scout car is easy, and only takes a few minutes. Printing the parts, however, is still a real commitment. Currently it's over 24 hours of print time, though I think it could be done in under 4 hours with a large-nozzle 3D print head like the E3D Volcano or 1.0mm Printrbot nozzle.

1. PRINT THE PARTS

Most of the Scout's parts can print on a 150mm × 150mm build surface, but the 2-part electronics housing is currently 156mm long. I recommend printing with PLA on a heated bed.

Download the parts from github.com/tlalexander/Flutter-Scout and get printing. There are detailed instructions there, but basically most parts can be 20%–30% infill, while gears, drive shafts, and steering pins should be 100% infill. Print the parts for the front end first, and note that the electronics bay and cover may need to be printed with a "brim" for extra adhesion.

2. TEST THE ELECTRONICS

Solder connectors onto the ESC for the motor and the battery, and cover all connections with heat shrink tubing (Figure Ⓐ). Lithium batteries have a lot of power, and short circuits can be dangerous.

Follow the instructions that came with the Flutter kit (or your radio) to attach the servo and speed controller, and then power everything up and make sure it works.

3. ASSEMBLE THE FRONT END

The front end consists of 2 wheel assemblies and the center frame. Place the bearings and drive shaft into each bearing assembly and slide in the shaft cover (Figure Ⓑ). Then for each wheel

assembly slide on 2 hub plates and a tire, and secure them with the wheel shaft clip (Figure Ⓒ).

Next, take the servo (with the "horn" removed) and slide it into its cavity in the frame, first forward at a 45° angle and then down, so that the front of the servo locks underneath the lip in the front of the cavity.

Secure the servo by sliding a steering pin through the small hole in the frame member above the servo flange, and flip over the assembly (Figure Ⓓ).

Attach the 6-lobed servo horn to the servo, capturing the steering rack beneath it (Figure Ⓔ). If your servo didn't come with a 6-lobed horn, a printable version is in the Github repository.

4. ATTACH THE WHEELS

First, wrap the rubber band around the top steering towers (Figure Ⓕ), and attach the 2

Taylor Alexander

Time Required:
1-2 Weekends
Cost:
$100–$200

Materials

» **Flutter Scout V2 3D print-ed parts** Download the files for printing at github.com/tlalexander/Flutter-Scout.

» **R/C transmitter and receiver** The new Flutter Vehicle Control Kit (flutter-wireless.com) has twice the range of standard R/C, and includes two Flutter Basic Boards, an Explorer Board, and a dual-joystick Remote Control Board with lithium battery.
  You can also use standard R/C gear.

» **Rubber band, thin**

» **Bearings, shielded, 12mm (6)** 6001Z type

» **Bearings, 8mm (2)** 608 type

» **Servomotor, metal gear** MG995 type

» **Brushless motor, 1,000kV** D2830-11 type

» **Propeller adapter, 3mm, collet type**

» **Electronic speed controller (ESC), 30A, with reverse** I recommend HobbyKing #HK-30A.

» **Bullet connectors, 3.5mm (3)** or motor extension wires, for connecting ESC to motor

» **Battery mating plug** for your ESC

» **Heat-shrink tubing**

» **Battery pack, lithium poly-mer (LiPo), 2S, 2,200mAh (1 or more)** roughly 104×34×17mm. I recom-mend HobbyKing's Zippy Flightmax #Z22002S20C.

» **Tires, foam rubber, 1:10 scale (4)** 7006 type, 95mm diameter, 36mm wide

» **Lithium battery charger/balancer** such as Turnigy E3

Tools

» **3D printer (optional)** with at least 156mm travel in one direction. If you don't have a printer, visit makezine.com/where-to-get-digital-fabrication-tool-access to find a machine or a printing service you can use. Or shop for great 3D printers at the Maker Shed, makershed.com/collections/3d-print-ing-fabrication.

» **Soldering iron**

» **Phillips screwdriver** for servo screw

» **Pliers**

NOTE: You may need to flex the frame slightly to get the shaft installed, but if it's too tight check to make sure the small hole in the gears is totally cleaned up and seating square against the shaft.

wheel assemblies using the short and long pins.

Then bring the front of the rubber band back over the steering towers as shown in Figure **G**. The rubber band isn't critical, but helps to keep the long steering pins from working their way out.

Power up the servo and make sure everything is centered. If the servo isn't centered when powered on, just unscrew the servo horn and adjust the position until the steering is straight.

5. ASSEMBLE THE REAR END

The rear frame holds the motor and the drive shafts. Start by collecting the pieces needed for the small, middle drive shaft.

Install the gears on both ends of the shaft with the large gear on the longer end. Place one 608 bearing on the end of the shaft by the large gear, and the other 608 bearing in the frame as shown in Figure **H**. Guide the bearing on the shaft into its hole at an angle, then straighten out the shaft and place the short end into the opposite bearing.

Secure the shaft with the 8mm shaft clip, aligning the flat inside the clip with the small notch in the middle drive shaft (Figure **I**).

Now you'll assemble the rear drive shaft and install it into the frame. Take the large, rear drive shaft and install the rear drive gear so that it mates with the rectangular section of the shaft, and then place 12mm bearings on both ends of the shaft. Lower the shaft into the frame so that the large rear gear mates with the small middle gear (Figure **J**), and then slide the bearings into their recesses.

Snap the 12mm shaft clip onto the rear shaft (Figure **K**).

6. INSTALL THE MOTOR

Assemble the motor, small drive gear, and quadcopter propeller adapter as shown in Figure **L**. These adapters were designed to turn only one direction, but our car spins the motor both directions, and the gear can loosen if it's not tight enough. So use the simple 3D-printed wrench to hold the gear while tightening the adapter with pliers. It should be pretty tight, but be careful as these aluminum propeller adapters can break with heavy tightening.

Now slide the motor into its hole in the rear frame with the wires facing up, and then lock it in place with the motor lock wedge (Figure **M**).

You should now have something that resembles a giant servomotor (Figure **N**).

7. FINAL BODY ASSEMBLY

Slide the front end assembly onto the large

dovetail in the rear frame. Then slide the electronics bay over the top, with the motor wires feeding into the side hole (Figure **O**).

Center the electronics bay over the rear frame, and drop in 2 short pins.

8. INSTALL THE ELECTRONICS

If you sourced the recommended parts, they just drop into place. Remove the power switch from the ESC, feed it through the power switch hole, and then plug it back into the ESC.

Next, plug the motor leads into the ESC. There are 3 leads and they can be plugged in to the ESC in any arrangement. If the motor spins backward, switch any 2 wires and you'll be good.

Hook up the servo wire and the speed controller to the radio, and then plug in the battery and neatly stow all the wires (Figure **P**).

9. BUILD THE REMOTE (OPTIONAL)

If you're using the Flutter boards, follow the instructions on the project page online at makezine.com/go/flutter-scout to print and assemble the remote control (Figure **Q**). (Otherwise you can use your standard R/C transmitter and receiver.)

10. CLOSE IT UP AND DRIVE!

Snap on the lid and you should be ready to roll. Go outside and test Scout's limits. Scout is fast and loves jumps. And if you break anything you can always print more parts!

We'd love to see people tweet pictures of their own Scout builds to @FlutterWireless. And if you want to talk to others about assembling Scout or about new design ideas, please visit community. flutterwireless.com. ●

O

P

Q

Get more step-by-step photos and share your build at makezine.com/go/flutter-scout

Semi-Automatic Coffee Roaster

Written by Larry Cotton

Time Required: A Weekend **Cost:** $100–$150

LARRY COTTON is a semi-retired power tool designer and part-time math instructor who loves music, computers, electronics, furniture design, birds, and his wife — not necessarily in that order.

1. Green bean hopper
2. Sephra chocolate-fountain auger
3. Roasting basket
4. Roasting basket speedometer
5. Dump beans switch
6. Coleman propane stove
7. Acrylic gear
8. Black and Decker AS6NG alkaline cordless screwdrivers (3)
9. Parallax HomeWork Board
10. Bean-cooling screen

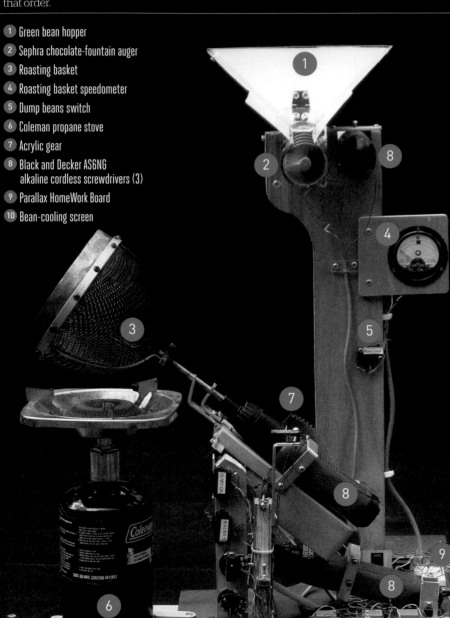

Three cordless screwdrivers and a microcontroller = perfect small-batch roasting

MY FIRST DIY COFFEE ROASTER FOUND ITS WAY INTO *MAKE:* **VOLUME 08.** Newly obsessed with fresh-roasted java, I called it the **Nirvana Machine** (makezine.com/go/nirvana-machine). It had — still has — two big advantages over others: it's easy to see the beans while they're roasting and it's portable.

However, there was always this lingering, increasingly urgent desire to improve it. The roaster required considerable intervention between batches: stopping the roasting by turning off the heat, dumping the beans, then reloading the basket and setting the temperature again on its little propane camp stove.

So I refined a new goal: create an easy-to-make, set-and-forget, continuous automatic coffee roaster. And I did. Several. Many. And they all failed in one way or the other — some quite spectacularly. But this one, a semi-automatic version, works beautifully.

Roasting is done with the basket's shaft (its spin-axis) at about 45°. The other basket positions were pretty obvious: horizontal for dumping the beans and vertical for loading them (via a repurposed auger for a chocolate fountain). One cordless screwdriver spins the basket, another moves it through the dump/reload arc (a worm gear drives a toothed acrylic gear), and a third drives the auger.

It's still prudent to keep an eye on the beans while they're roasting. But the rest of the process — dumping and reloading — is automatic: just press one button when the beans are roasted and a microprocessor controls the rest. I happened to have an ancient Parallax HomeWork board on hand, but an Arduino would work as well. Salud! ⊘

Larry Cotton

Read the complete build saga and see how it all works at makezine.com/go/semi-auto-coffee-roaster.

PROJECTS

1 2 3 Realistic Duct Tape Rose

Written and Photographed by Jason Poel Smith

SURPRISE YOUR SWEETHEART WITH A REALISTIC, HANDMADE ROSE crafted from an unexpected material — duct tape!

1. MAKE THE PETALS

Cut off a 1' length of wire and a 4" piece of red duct tape. Stick the wire to the tape so that they overlap by 1½". Then fold the tape over the end of the wire and stick the 2 sides together, leaving about 1" of sticky tape exposed.

Cut the petal to shape by rounding off the corners, and color in the cut edges with markers. Make about 10 petals.

2. BUILD THE BUD

Take the first petal and loosely roll it into a tube. Then softly wrap each additional petal around the previous one.

As you go, bend a gentle curl into each petal to give it the shape of an actual rose.

3. MAKE THE STEM

Build the stem by twisting the wires together and wrapping them in green duct tape.

To make the sepals, take several pieces of green duct tape and fold one end over, similar to the way the petals were made. Cut out 5 small triangles with the exposed tape at the base, and color in the cut edges. Then attach these to the base of the flower. Your duct tape rose is complete. ⊘

JASON POEL SMITH makes the "DIY Hacks and How Tos" project video series on *Make:*. He's a lifelong student of all forms of making, from electronics to crafts and everything in between.

You will need:

- » Duct tape, red, ~3-4' per rose
- » Duct tape, green, ~2-3' per rose
- » Floral wire or other stiff wire
- » Markers, red and green
- » Scissors
- » Wire cutters (optional)

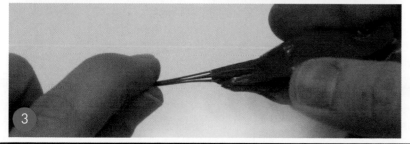

See the how-to video and more photos and tips at makezine. com/go/duct-tape-rose

Hep Svadja

Henry Bessemer and the Age of Steel

Written by William Gurstelle

Learn how to heat-treat and toughen the metal that built our world: carbon steel

WILLIAM GURSTELLE is a contributing editor of *Make:* magazine. His latest book, *Defending Your Castle*, is available at all fine bookstores.

Time Required:
An Afternoon

Cost:
$5–$10

Materials

» Music wire, ³/₃₂" diameter, 36" length
» Heat-shrink tubing, ³/₁₆" diameter, 8" length
» Rubber bands
» Soybean oil, 1 quart or liter

Tools

» Propane torch
» Steel pipe, nominal ¾", about 12" long Note that ¾" steel pipe has an outside diameter of about 1".
» Pliers, needlenose, large
» Pliers, slip joint
» Gloves, heavy
» Safety glasses
» Hot plate
» Candy thermometer
» Pot with lid
» Steel wool
» Bowl of water
» Hacksaw
» Vise

AT THE START OF THE INDUSTRIAL REVOLUTION, THE STATE OF THE ART IN IRON METALLURGY WAS THE PUDDLING FURNACE. Ironworkers loaded crude "pig iron" ingots into the furnace and then continually stirred the molten metal through a small hole. Puddling was a hot and difficult task, and an experienced puddler was considered a highly skilled craftsman.

As the puddler stirred, solid chunks of refined iron would appear in the liquid mass. He'd gather these and work them under a forge hammer to yield a slab of hot wrought iron which, after all that work, still lacked the strength or utility of steel.

Steel is iron alloyed with carbon and it's superior to plain iron in every way. But prior to 1856 there was no practical way to control the percentage of carbon in iron, so there was no way to manufacture steel at a price industry could afford.

The railroads were booming in the mid-19th century, but their wrought iron rails were too soft — on busy stretches the rails had to be replaced every 6 to 8 weeks. Steel rails would be far more durable, but they were too expensive.

Then, a smart fellow named Henry Bessemer came on the scene. Bessemer, an English engineer and metallurgist, would receive 129 patents in a variety of engineering disciplines. But the invention for which he was knighted and the one that made him rich was the one that involved turning iron and coke (carbon) into steel.

While looking for a way to strengthen cannon barrels, Bessemer discovered that the carbon dissolved within molten pig iron unites readily with oxygen. Knowing this, he determined that if he could blast a jet of air through the molten pig iron, then he could convert it into much stronger alloy steel by accurately controlling its carbon content.

Bessemer built an experimental furnace at his laboratory in London, with a high-temperature heating chamber 4 feet high, and a 12-horse-power steam engine to run the air injector. When the pig iron in the chamber liquefied and he turned on the blower, a fireball erupted from the top. But when he sluiced the molten metal into ingot molds, he gazed with delight upon the "limpid stream of incandescent malleable iron almost too brilliant for the eye to rest upon." Bessemer had found a way to make cheap steel.

In 1856 Bessemer designed what he called a converter (Figures Ⓐ and Ⓑ), a large, napiform

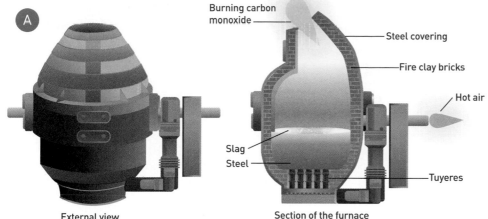

External view

Section of the furnace

Burning carbon monoxide

Steel covering

Fire clay bricks

Hot air

Slag

Steel

Tuyeres

James Burke

Alfred T. Palmer

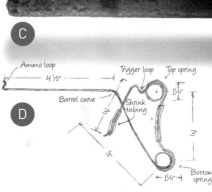

Ammo loop
4½"

Barrel curve

Shrink tubing

Trigger loop

Top spring

1¼"

3"

3"

Bottom spring

1¼"

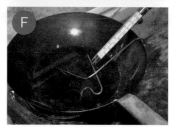

Hep Svadja

(turnip-shaped) receptacle with holes ("tuyeres") at the bottom where pumps could inject compressed air. Bessemer filled his converter with molten pig iron, blew compressed air through it, and found that the pig iron was indeed emptied of excess carbon and silicon in minutes. From that time onward, affordable carbon steel could be made abundantly. The Age of Steel had begun.

MAKE A STEEL RUBBER BAND GUN

Steel is such an important industrial metal because it is strong yet malleable. Further, steel can be heat-treated to make it harder or softer, flexible or stiff, ductile or brittle, depending on the application.

In this project, you'll heat-treat steel to make a rubber-band shooting gun: first you'll *anneal* it to make it soft and malleable, then cut and form it to shape, *quench* it to harden it, and finally *temper* it to give it the tough, springy characteristics your gun requires.

1. Don safety glasses and heavy gloves.

2. Place one end of the music wire in the vise. At this point, the wire is very stiff and difficult to shape. Light the torch and heat the wire to cherry red (Figure **C**), moving slowly down the wire until the entire wire has been heat-treated. (Cherry red is a fairly dull color. If it glows bright orange, you've heated the metal too much.) The heating and subsequent air-cooling of the wire is a process called annealing, which makes the stiff music wire soft and malleable.

3. Place the steel pipe in the vise and form the cooled wire around the pipe to make the round top

and bottom springs, as shown in Figure **D**.

4. Use the pliers and vise to bend the ammo loop, the trigger loop, and the barrel curve into shape as shown in Figure D. Cut off any excess wire with a hacksaw or rotary cutoff tool.

5. Heat the top and bottom springs with the torch until they glow cherry red (Figure **E**). Once hot, plunge them into a bowl of water to quench them. Dry the springs and clean off the scale (iron oxide) with steel wool. Take care — after quenching, the steel in the spring is very hard and extremely brittle. Bending it even a little will cause it to break.

6. Pour oil into the pot and heat on the hot plate to 400°F, using the thermometer to check the temperature. Use caution here — the hot oil may smoke, so do this outdoors or keep a window open and ventilate the room.

7. Reheat the springs with the torch until they glow cherry red. Stop heating and allow them to cool just until the red color is completely gone, and then plunge the springs into the hot oil (Figure **F**). Turn off the hot plate and allow the oil and the springs to cool slowly to room temperature. This very slow cooling process will reduce hardness and restore the tough, springy temper to the steel.

8. Clean off the oil and thread 3 pieces of heat-shrink tubing over the handle and trigger as shown, and then shrink the tubing using a match or the hot plate.

If the heat-treating process worked properly, you now have a springy steel rubber band shooter (Figure **G**). Load up and pull the trigger! ⊘

SAFETY NOTES
Use extreme care when handling the hot oil and wire. The smoke point of soybean oil is very high, around 450°F, and even then doesn't ignite readily. Still, take care to:
» Use an electric hot plate, not an open flame
» Not let the oil heat up past the flash point
» Have a lid handy to cover your pot in the unlikely event of a problem
» Keep a grease-fire capable extinguisher close by.

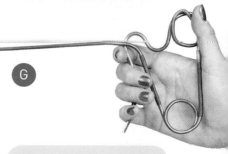

See more photos and share your build at makezine.com/go/steel-rubber-band-gun.

Time Required: 1-2 Hours Cost: $10-$20

The Greenest
Delay Timer

Build a clever timer circuit that draws no current at all between cycles. What could be greener?

Written by Charles Platt

CHARLES PLATT is the author of *Make: Electronics*, an introductory guide for all ages, and its sequel, *Make: More Electronics*. He also wrote Volumes One and Two of the *Encyclopedia of Electronic Components*. Volume Three is in preparation. makershed.com/platt

Materials

» **Resistors, ¼W: 100Ω (1), 470Ω (1), 1kΩ (1), 10kΩ (3), and 1MΩ (1)**
» **Capacitors: 0.01µF (2), 0.068µF (1), 2.2µF (1), 10µF (2), and 1,000µF (1)**
» **LED**
» **Tactile switch, momentary pushbutton** Alps Electric #SKQNAED010 or similar
» **555 timer IC chips, TTL type (2)**
» **Relay, DPDT, 9VDC** Omron #G5V-2-H1 DC9 or similar
» **Speaker, 2" or 3"**

Tools

» **Solderless breadboard**
» **Wire cutters/strippers**

THERE'S NEVER A SHORTAGE OF APPLICATIONS FOR A DELAY TIMER.

Outdoors, you may want to switch on a light for just 1 minute while you get from your house to your car. In the kitchen, you wait for a beep when the food's cooked. You might want to run a doorbell for a significant interval so you'll be sure to hear it — or switch off a heat lamp in the bathroom in case you forget.

The problem I have with delay timers is the hassle of powering them. I don't necessarily want to place one near a power outlet, but I don't want to worry about replacing batteries, either. CMOS devices use only a tiny trickle of current while they're sitting there waiting patiently for you to activate them, but still, it bothers me that they suck electricity while doing nothing.

I decided to design the greenest possible delay timer — one that uses no current at all between one cycle and the next. Zero power! If it was battery-powered, a fresh 9-volt battery could last 4 or 5 years.

A ZERO POWER CIRCUIT

The circuit I came up with is unusual but simple. When you press a button, it switches on a 555 timer chip, which activates a relay for a fixed interval. At the end of the interval, the relay switches off the timer, and the timer switches off the relay. Does that sound implausible? Take a look at the schematic in Figure Ⓐ, which I've laid out so that it's easily transferred to a breadboard.

When the button is pressed, it supplies power to pin 8 of the timer. Because the button is only switching on the timer, not triggering it, I had to add a 10µF capacitor to pass an initial low state through a 1K resistor to trigger the timer on pin 2, after which a 10K resistor maintains pin 2 at a high state.

The timer is wired in monostable (one-shot) mode, to emit a single delay pulse. The pulse goes out of pin 3 to the relay, which I have drawn to show its internal contacts. When the right-hand contacts close, they feed power back to pin 8 of the timer. So now the relay is powering the timer, and will continue to do so after the button is released, because the output pulse from the timer is still running the relay. The timer and the relay are sustaining each other — until the output pulse from the timer ends. Then the relay contacts open, which switches off the timer. At that point, we have zero power consumption, because the circuit between the positive bus and the negative bus is completely open.

Naturally you'll need to adjust the output pulse of the timer to suit your application. Here I've chosen component values for a pulse lasting about 2 seconds. For a longer pulse, search online for "calculate 555 duration" and you'll find sites that tell you what component values to choose. I suggest you keep the 1M resistor and increase the value of the 2.2µF capacitor. A 56µF capacitor should create a pulse lasting about 1 minute. A 1,000µF capacitor should provide 18 minutes.

The left-hand contacts in the relay can be used to switch anything you like, within limits set by the manufacturer's datasheet (typically, a couple of amps). A high-wattage light running on 115VAC would be acceptable.

ADDING A BEEPER

But what if you want the timer to run in "kitchen mode," so that it beeps when the timing interval is complete? Can this be done while still maintaining zero power consumption in the dormant state?

Yes, you just need a second timer powered briefly by a large capacitor. In Figure Ⓑ I have extended the previous circuit downward. The upper left contact of the relay now receives power

Charles Platt

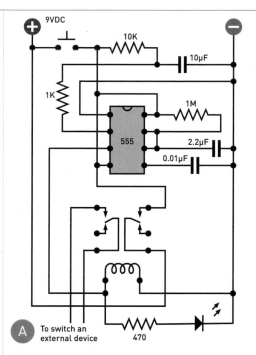

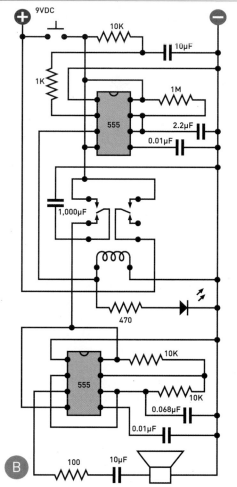

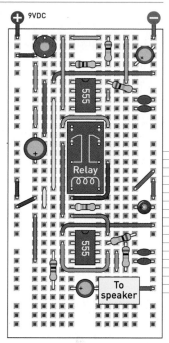

A To switch an external device

Figure A. The greenest delay timer in its simplest configuration, allowing you to attach the device of your choice to the contacts of the relay that close when it is activated.

B

Figure B. The extended greenest delay timer, including a chime when it reaches the end of its cycle.

Resistors, ¼ watt:
100 ohm (1)
10K (3)
1K (1)
470 ohm (1)
1M (1)

Capacitors:
0.01µF (2)
0.068µF (1)
2.2µF (1)
10µF (2)
1,000µF (1)

Other components:
LED
Tactile switch
555 timers, TTL type (2)
Relay, DPDT, 9VDC
Speaker, 2" or 3"

C

Figure C. Breadboard layout for the extended greenest delay timer.

from the upper right contact of the relay, and a 1,000µF capacitor charges from the left-hand contacts. Then when the timer pulse ends, the contacts relax and this capacitor discharges into pin 8 of a second 555 timer, which is wired to create a musical chime. The chime lasts about 1 second, until the capacitor exhausts itself. After that, once again, the circuit draws zero power.

Figure C shows a breadboard layout of the full circuit. You can adjust the duration of the chime by increasing or decreasing the value of the 1,000µF capacitor, and you can change the pitch of the chime by adjusting the resistor and capacitor values associated with the second 555 timer. Again, you can search online for appropriate values.

ADJUSTING THE DELAY

Of course in a kitchen-type timer, you'll want to be able to choose a variety of possible delay values. Can we do that and still retain our zero-power status? Don't we need to add a counter and a numeric display?

No, certainly not — so long as you're willing to be retro and think analog. Simply replace the 1M resistor with a 1M potentiometer and 10K resistor in series. (This resistor prevents the pot from reducing the resistance all the way to zero.) Now you can adjust the delay simply by turning the potentiometer.

You'll have to calibrate it by trial and error, but that shouldn't take long. And think of the payoff. When a guest claims to have a green lifestyle, you point smugly to your timing device and say, "That gadget is using absolutely no power whatsoever." How could anything be greener than that?

THE GREENEST ENCLOSURE

How to fabricate an enclosure for the greenest timer? Normally I'd use ABS plastic, but for this project, that would be heretical. Out in the yard I found the answer: an old pine 2×4, seasoned by years of harsh Arizona weather. For an impressive pushbutton, I used a soft rubber bulb syringe (Figure D), intended for aspirating mucus from a baby's nose.

The final enclosure looks as if it might satisfy the most sustainable community. See how I made it at makezine.com/go/greenest-delay-timer.

See more photos and share your build at makezine.com/go/greenest-delay-timer.

Figure D. A soft rubber bulb syringe, intended for the nostrils of infants, makes for an impressive pushbutton.

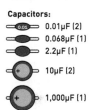

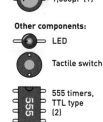

Use LEDs to Track Night-Launched Projectiles

Written by Forrest M. Mims III

MODEL ROCKETS AND TOY PROJECTILES SENT ALOFT BY RUBBER BANDS OR SLINGSHOTS ARE LOADS OF FUN. While model rockets are best for high altitude, you can have plenty of fun launching inexpensive projectiles equipped with a steady-state or flashing LED and flown at night. You can stage competitions to see how far various designs can travel. And you can do some real science by recording their flights with a time exposure camera to study flight stability, spin, and speed. Let's get started with an ultra-simple projectile launched with a slingshot.

SLINGSHOT MICRO-LIGHT ROCKET
Hobby Lobby stores sell a package of a dozen miniature white LEDs installed in tiny metal cylinders equipped with a battery and rotating switch. These bullet-shaped lights are ideal for use as tiny inertia projectiles launched from a slingshot at night. They can be launched as-is or stabilized by taping one end of a broom straw, 6" long or so, to the LED fixture so that it resembles a miniature bottle rocket. Switch on an LED (or two) and place it in a slingshot pocket (Figure 1). If a broom straw is attached, it should point straight up so it won't hit the slingshot fork. Then fire the projectile into the air from a grassy field.

Air resistance has very different effects on these tiny projectiles. The unmodified LED will tumble in flight, while a broom straw version will usually provide a more stable flight. However, the bare LED sometimes reaches a higher altitude than the broom straw-equipped LED. This can be seen in the time exposure in Figure 2: Launched simultaneously by the same slingshot, a bare LED reached 102.6 feet and a straw-stabilized LED only 90.8 feet. The altitude was determined by using the known height of the adjacent tower illuminated by a flashlight as a reference.

You can have lots of fun with these micro rockets while doing some real science. For example, what's the minimum length of the straw for stable flight? What's the maximum distance achieved by launching micro-rockets at an angle? How high

FORREST M. MIMS III (forrestmims.org), an amateur scientist and Rolex Award winner, was named by *Discover* magazine as one of the "50 Best Brains in Science." His books have sold more than 7 million copies.

2

can they travel? Why does the bare LED travel as high or higher than the straw-stabilized LED, and why is its tumble so regular?

You can also organize competitions to see whose micro rocket covers the most distance or lands closest to a target on the ground. Paint the LEDs with different colors and leave them glowing in the field until all the competitors have completed their flights.

MORE POWERFUL INERTIA ROCKETS

During a recent visit to the White Sands Missile Range Museum, I purchased a package of three Sky Rockets made by Monkey Business Sports. These "High Flying, Hand Launched, Foam Rockets" are sent on their way by a handheld elastic launcher. If you flick your wrist forward while releasing the rocket, it can reach an altitude of 300 feet.

It's easy to equip a Sky Rocket with a tracking light. Tape the leads of a white LED across opposite sides of a lithium coin cell so that the lead on the flat side of the LED is on the negative side of the cell, and tape this assembly to the back end of a rocket (Figure 3). (Optionally, you can add a simple on-off switch using a paper tab, as shown in Step 3 at makezine.com/go/extreme-led-throwies.)

To study roll rate, insert a Hobby Lobby LED into a hole made in the side of the rocket at its center of gravity (balance point) and secure it in place with clear tape (Figure 4). The time exposure in Figure 5 shows a typical flight, with the rocket's rotation clearly indicated by streaks of light. The streaks are longest during the early stage of flight when the missile is moving fastest. The ground streaks in the photo are from my flashlight as I walked to the launch site.

For a tracking light that flashes, I've not found anything better than Coghlan's Brite Strike APALS, an ultra-thin, lightweight rectangle measuring 1"×2" that is easily stuck to the side of a rocket and secured with clear tape. The time exposure in Figure 6 shows the apogee of a flight neatly outlined by this red flasher.

PHOTOGRAPHING NIGHT-LAUNCHED MISSILES

Digital cameras are ideal for recording the flight path of night-launched rockets and inertia projectiles equipped with a tracking light. The flight paths in Figures 2, 5, and 6 were photographed with a Canon 7D set for an ISO of 6400. The lens was a 20mm–40mm wide angle. The camera was mounted on a tripod 150 feet from the launch site.

For early flights I set the exposure time to 30 seconds, tripped the shutter, ran to the launch site, and launched the rocket using an elastic-powered hand launcher or slingshot. After falling down during one night session, I began triggering the camera with a radio-controlled actuator, which greatly simplified the launch protocol. After the camera was set for a 10-second exposure, I walked to the launch site, pressed the transmitter button and launched the rocket.

The altitude reached by your rockets can be determined by including an illuminated object of known height in each photo. Dark nights with few or no clouds are best.

GOING FURTHER

The sky's the limit with this project. You can easily build your own inertia projectiles using paper or plastic tubes equipped with paper fins and a foam nose from a toy missile or a miniature paint roller. I've built and launched simple rockets using a homemade launcher I made by attaching a loop of neoprene tubing to a handle. Night tracking would also be fun with water rockets (makezine.com/go/soda-bottle-rocket), compressed-air rockets (makezine.com/go/high-pressure-foam-rocket), and other DIY rockets and projectiles.

SAFETY PRECAUTIONS

As with any project involving projectiles, use common sense. Children should be supervised. Rockets should never be launched toward people or structures and should always be launched from a field. Inspect the field during daylight to check for potential safety hazards.

Wear clear safety glasses when launching elastically propelled rockets. During night launches, all participants should have a flashlight. I wear a headlamp to keep my hands free. ◐

Follow this project and share your night launch experiments at makezine.com/go/tracking-night-launched-projectiles.

Forrest M. Mims III

Time Required:
An Evening
Cost:
$10–$35

Tools

» **Toy rocket, elastic-launched** such as Sky Rockets by Monkey Business Sports
» **Slingshot**
» **LED, self-contained with battery assembly** Hobby Lobby #101675 or similar
» **LED, white**
» **Lithium coin cell, CR2325**
» **Light strips, adhesive** Coghlan's Brite Strike APALS, from a hobby shop
» **Adhesive tape, clear**
» **Camera** with bulb mode or time-exposure setting
» **Tripod**
» **Flashlight**
» **Safety glasses**

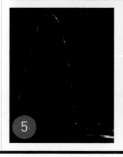

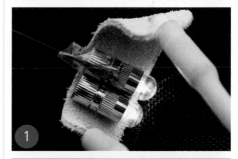

Water Balloon Cannon

Written by Jason Fabbri

Build a simple PVC air cannon to help you win the water wars

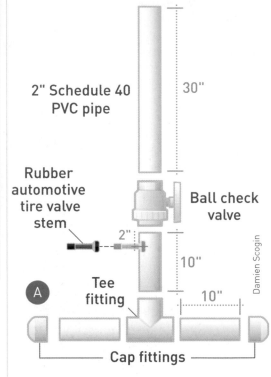

2" Schedule 40 PVC pipe — 30"

Rubber automotive tire valve stem

Ball check valve

2"

10"

10"

Tee fitting

Cap fittings

(A)

Damien Scogin

JASON FABBRI is a software engineer, father of three boys, and avid builder of wacky things. When he's not conceiving crazy ideas he can be found wrenching on a vintage motorcycle.

Time Required: 20–30 Minutes
Cost: $20–$25

Materials

» **PVC pipe, 2", Schedule 40 or 80, 5' length** Don't use Schedule 20, it's too thin.
» **PVC pipe fittings, 2": tee (1), caps (2), and ball check valve (1)**
» **PVC glue**
» **Automotive tire valve stem, rubber** not steel

Tools

» **Saw**
» **Sandpaper or file**
» **Drill and bits**
» **Bike pump**

TELL ME IF THIS SOUNDS FAMILIAR: THE KIDS ON THE STREET START A WATER BALLOON WAR, THEN RUN JUST OUT OF YOUR REACH. Well, I'm here to help. Let's add a new weapon to your water fight arsenal: a pneumatic water balloon cannon. This is an easy project to put together. For about $20 and 20 minutes of time you can return fire across the street, down the street, or even to the next block.

You will strike fear into all those who seek to soak you.

1. CUT THE PIPE

Cut the PVC pipe into one length of 30" and 3 lengths of 10". Use sandpaper or a file to remove any burrs.

2. INSTALL THE VALVE STEM

Mark a 10" pipe about 2" from one end, and drill a hole for the valve. Start too small, and slowly enlarge it using larger bits or a file; you want a very tight fit for an airtight seal.

Insert the valve stem from inside the pipe, wiggling it through until it seals securely.

3. GLUE IT UP

Assemble the cannon as shown: first the tee and caps, then the ball valve, and finally the barrel (Figure (A)). For a good seal, apply glue liberally and give each joint ¼ turn.

4. FIRE IN THE HOLE!

To fire water balloons, follow these steps:
» Close your check valve. If you leave it open, your balloon will fall into the "air tank" area never to be seen again.
» Put the water hose into the barrel and fill it about 3"–4". This water is crucial — it prevents the balloon from exploding on exit.
» Fill a water balloon almost exactly the diameter of the barrel, about 2". An overfilled balloon won't travel well.
» Fill the air tank. Attach an air compressor or bicycle pump to the valve stem. Do not overfill. I suggest 20–30psi.
» Make sure you're not aiming at anyone directly (straight up is a good first test) and open the valve with a very quick motion. *Pooof!* There goes your balloon, followed by a trail of water.

There you have it. A simple yet effective water balloon cannon that will be the envy of every kid on the block. Happy water wars! ◐

See step-by-step photos and video at makezine.com/go/water-balloon-cannon.

PROJECTS

1 2 3 Custom Soda Cooler

Written by Jason Poel Smith ■ Illustrated by Andrew J. Nilsen

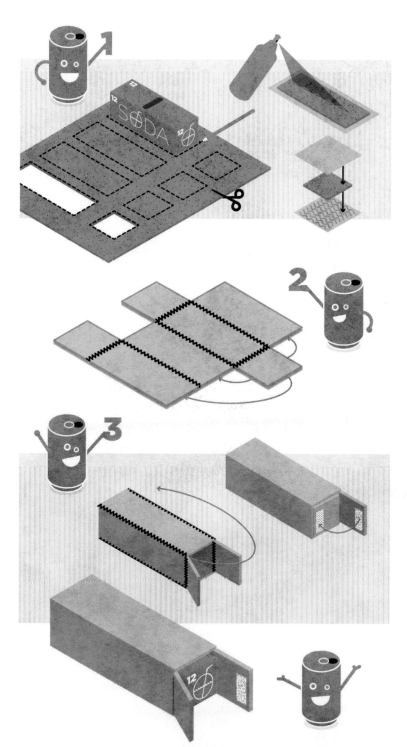

SUMMER IS HOT. SO WHY NOT MAKE A CUSTOM COOLER to keep your drinks cold! Here's how to make a flexible cooler that perfectly fits a 12-pack of soda. Take the whole pack out of the fridge, slip it into your custom cooler, and go.

1. MAKE THE SIDES

Start by determining the shape of your cooler. » Trace the outline of each side onto the foam. » Then cut out each outline.

Apply a generous amount of spray glue to one sheet of vinyl. » Then stick on the foam cutouts, spacing them out by at least 1". » Apply spray glue to a second sheet of vinyl and stick that on top. » Dry overnight, then cut out.

2. ASSEMBLE THE COOLER BODY

Lay out the sections as they will be assembled, and glue or sew the pieces together. » After attaching the sides together, trim the seams to no more than ½".

3. INVERT AND ADD A CLOSURE

Carefully turn your cooler inside out, so that all the seams are on the inside. If you're worried about breaking a weak seam, you can secure your seams with binder clips before inverting the cooler.

Finally, add a zipper or velcro that will hold the opening of the cooler closed. Attach this by gluing or sewing it in place. You're ready to chill. ◗

See step-by-step photos and video and share your cooler designs at makezine.com/go/custom-cooler.

You will need:

» **Vinyl sheet** I used a vinyl tablecloth.
» **Flexible foam sheet** such as craft foam
» **Sewing machine** or spray glue
» **Velcro or a zipper**

JASON POEL SMITH makes the "DIY Hacks and How Tos" video series on *Make*. He is a lifelong student of all forms of making, and his projects range from electronics to crafts and everything in between.

PINEWOOD DERBY!

READY... SET... ROLL!!!

MATERIALS

- -1/4" X 16" PLYWOOD 8' LONG (2) FOR BASE
- -1/4" X 1.5" WOODEN TRIM 8'4" LONG (4) FOR TRACK
- -1/4" X 1/5" WOODEN TRIM 7'8" LONG (4) FOR TRACK
- -1/4" X 1/5" WOODEN TRIM 16" LONG (8) FOR SUPPORT
- -1X4 LUMBER 16" LONG (4) FOR SUPPORT BEAMS
- -2X2 LUMBER 16" LONG (1) FOR LADDER SUPPORT
- -1/2" PVC PIPE 30" LONG WITH (2) ELBOW JOINTS AND (2) END CAPS
- -1" WOOD SCREWS
- -1/2" FINISHING NAILS

BUILDING THE STARTER:

END CAP →

CREATE THE **STARTER** HANDLE WITH PVC 1/2" PIPE AND FITTINGS.

ELBOW JOINTS

3"

9"

8"

1.5"

18"

END CAP

MARK A 1-1/2" LINE DOWN THE CENTER OF EACH TRACK, STARTING 8" FROM THE TOP. DRILL 5/16" HOLE AT THE TOP AND BOTTOM OF LINE. USING A KEYHOLE SAW, CUT THROUGH UNTIL THE WOOD IS REMOVED.

USE 1/2" **PIPE CLAMPS** WITH 3/4" MACHINE SCREWS AND BOLTS TO SECURE PVC STARTER TO UNDERSIDE OF THE TRACK.

DRILL 2" MACHINE SCREWS ALL THE WAY UNTIL THEY HIT THE OPPOSITE WALL OF THE PIPE, MAKING SURE THAT THEY **LINE UP EXACTLY** WITH EACH OTHER AND CAN REST ON EDGE OF THE HOLE.

WHEN **LOADING** THE CARS, KEEP THE PVC HANDLE DOWN TO COUNTERBALANCE THE WEIGHT OF THE CARS.

PUSH THE HANDLE **FORWARD**, PIVOTING THE SCREWS DOWN AND RELEASING THE CARS.

STEP LADDER SUPPORT:

A **STEP LADDER** IS USED TO SUPPORT THE TRACK AND ALLOWS KIDS TO EASILY CLIMB AND REACH THE PVC PIPE STARTER.

2X2 LUMBER **LOCKS** INTO LADDER PREVENTING **SLIPPAGE** →

1" X 4" X 16" **SUPPORT BEAM**

BUILDING THE TRACK:

GLUE WOOD TRACK DOWN AND REINFORCE WITH 1/2" FINISHING NAILS.

EXTRA BEAMS FOR SUPPORT CAN BE ADDED TO THE BOTTOM OF THE TRACK. THIS CAN ALSO PREVENT NAILS FROM GOING ALL THE WAY THROUGH.

THE TRACKS OVERLAP THE SECOND PLYWOOD BOARD AND ARE SECURED WITH WOOD SCREWS. THE SECOND BOARD IS SUPPORTED BY A 16" X 1" X 4" WOOD BASE HELPING TO SECURE THE TRACK.

THE 8 FT BOARDS ARE CONNECTED USING 1" WOOD SCREWS. THE SCREWS CAN BE REMOVED FOR EASY STORAGE.

THE FORCE OF GRAVITY PULLS THE CAR DOWN THE TRACK. THE SPEED COMES FROM WEIGHT, WEIGHT DISTRIBUTION, AND AERODYNAMICS. EXPERIMENT WITH THE SHAPE OF YOUR CAR!

THIS IS AWESOME! LET'S RACE AGAIN!

BUILDING THE CAR:

MATERIALS

2X2 LUMBER 6" LONG CUT AND SHAPED

3/16" X 1-1/2" FENDER WASHERS (4)

1/4"
1"

10 X 3/4" SCREWS (4)

DRILL A STARTER HOLE TO HELP ALIGN SCREWS.

DRILL SCREWS IN LEAVING ROOM FOR WHEELS TO SPIN FREELY.

GET MORE AT...

HOWTOONS.com

PINEWOOD DERBY CAR RULES:

1. WIDTH SHALL NOT EXCEED 2-3/4 INCHES.
2. LENTH SHALL NOT EXCEED 7 INCHES.
3. WEIGHT SHALL NOT EXCEED 5 OUNCES.
4. CARS MUST BE SHAPED FROM THE ORIGINAL BLOCK OF WOOD.

*THESE RULES MAY NOT MEET NATIONAL STANDARDS.

Pixel **PALS**

Learn how to solder Chip and his Pixel Power base, then plug him into an Arduino Written by Paul Gentile

Time Required: 30-60 Minutes   **Cost: $16**

Materials

» **Pixel Pals: Chip with Pixel Power Soldering Kit** Maker Shed #MKSS01, makershed.com. Includes printed circuit boards (2), LEDs (8), resistors (2), headers (male and female), micro pushbuttons (2), slide switch, coin cell battery and holder

Tools

» **Soldering iron and 60/40 solder**
» **Tip cleaner**
» **Pliers**
» **Safety glasses**
» **Masking tape**

PAUL GENTILE has been making stuff his whole life, from model trains to multicopters. He and fellow maker Jean Consorti founded Soldering Sunday (solderingsunday.com) to help Makers of all ages and skills create what they're passionate about.

WHETHER HE'S STANDING ON HIS PIXEL POWER BOARD OR ON AN ARDUINO, CHIP'S BIG LED EYES DRAW YOU IN, and his large size makes him easy for makers of all ages and skills to build. He's a great tool for learning to solder and for getting started with Arduino.

1. ASSEMBLE CHIP

Place the short pins of the header through Chip's feet on the front of the PCB, then turn Chip over. You can tape Chip to your work surface to keep him from moving around. Solder the short pins on the backside of the PCB.

Insert the resistors — either direction there's no (+) or (−) side — so they sit flush on Chip's chest. Spread the leads to hold them in place, solder them on the backside, and clip the leads just above the solder ball, not flush with the board.

Now add the LED eyes: Choose your favorite color(s), then insert the long leg (anode, positive) into the square pad and the short leg (cathode, negative) into the round pad. Solder and clip the LED leads.

2. ASSEMBLE THE PIXEL POWER BASE

Tape the slide switch in place — orientation doesn't matter. Turn over and solder. Insert the left and right pushbuttons and make sure they're flush. Solder in place.

Use tape to hold the 10-pin female header flush and square. First solder the end pins to secure the header, then complete the job by soldering each pin. Insert the battery holder and solder it in place.

3. PLAY!

You're done! Plug Chip into the Pixel Power Base and insert the battery. Now you can blink Chip's eyes by playing with the buttons.

You can also control Chip's eyes with an Arduino. Find the instructions and upload the code from the project page online at makezine/go/pixel-pal-chip.

We hope that you enjoy Chip and have many adventures together. ◉

See step-by-step photos and video, and share your photos of Chip on the project page makezine.com/go/pixel-pal-chip and Twitter @SolderingSunday — we love to see all the friends Chip has made.

Paul Gentile

Hep Svajda

PROJECTS

Toy Inventor's Notebook
FOOTSTEP SAND STAMPERS
Invented and drawn by Bob Knetzger

Time Required: 1 Day   Cost: $5-$10

THIS!   NOT THIS!

HERE'S A FUN AND EASY PROJECT FOR SUMMER: Hack a pair of flip-flops so they make designs in the sand as you walk. Just add a bit of Sugru to your sandals and start stamping!

First, clean the bottom of your flip-flops thoroughly with some vinegar water to get rid of any grease or loose dirt. Dry off the soles completely so the Sugru will stick and stay on.

Open the pack and knead the Sugru until it's pliant and soft, then form the shapes you want. Make shapes by first rolling out a small "snake," then press it into the sole of the flip-flop. Spread the bottom out and press down firmly for good adhesion. Don't make tiny details — you won't really see them in the sand. Think simple and bold shapes. Make letters, numbers, emoji, symbols, and designs but whatever you make, remember to make it backward, like on a rubber stamp!

Press down all around the edges of the Sugru. For best results, form your designs with a triangular cross-section, as shown, to maximize adhesion and eliminate any undercuts. Let the Sugru cure for 24 hours, then go sand stamping!

Works best in moist sand. I got great results even with a tired old pair of flip-flops and a packet of expired Sugru.

What design will you make? Happy stamping! ◗

See photos of the Footstep Sand Stampers in action, and share your designs at makezine.com/go/footstep-sand-stampers.

Sugru is available in the Maker Shed: makezine.com/go/sugru

TOOLBOX

Rockler Bandy Clamp

$19.99 a pair, $50 for 3 pairs :
rockler.com

As a professional photographer, clamps are a huge part of my kit, both in and out of the studio. Rockler Bandy Clamps hold up like a champ, from clipping cables to poles and holding gear in place, to keeping my seamless backdrop smoothly pinned to the table. The spring-loaded action is strong, but still easy to open for those times when you need to clamp something one-handed. As someone with smaller hands, I often have trouble manipulating larger clamps, but the rubber-molded grip of the Rockler Clamps makes one-handed use easy, and the all-nylon/fiberglass construction reduces the weight of my kit.

The no-mar clamp pads offer a sure grip, but are also padded to keep from scarring the surface of the clamped object, even for delicate items like paper backdrops. The pads even swivel so that the clamps can be used on items with odd angles, or to clamp to things like light poles where normal clamps would struggle with the rounded surface. The 2-inch maximum reach of the jaw means I can use them to clamp backdrops to thicker tabletops without having to carry around a bunch of redundant C-clamps.

— *Hep Svadja*

Hep Svadja

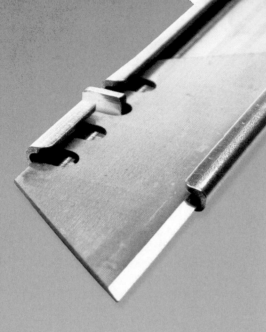

KLEIN DUAL LASER IR THERMOMETER

$130 : kleintools.com

Infrared laser temperature guns, like many tools, are greatly affected by price. The cheap options tend to have narrow temperature ranges and low spread-to-distance ratios — meaning the further you get from the object or area you're trying to measure, the larger the space that is being evaluated for its temperature becomes.

The Klein IR2000A is not one of those cheap options, and it shows quickly.

The most noticeable trait of the IR2000A is its high-temperature capacity of 1,022°F, allowing for measurement of everything from scalding hot water pipes to the hottest wood-fired pizza ovens. And unlike most thermometer guns, it also has a K-type thermocouple input that allows you to use an included physical probe to measure even more extreme environments, up to 1,400°F.

A 12:1 spot ratio means that you can pinpoint a 1"-diameter sample area when measuring an object from 12 inches away — a step up from the 8:1 spread seen on cheaper units. And dual lasers that widen with distance give a clear indication of the space that is being measured.

Again, it's not cheap, but the combination of range, precision, and Klein's rugged casing (it's rated for falls from over 6') make this one of my more trusted and regularly used tools.

— *Mike Senese*

SCREWPOP UTILITY KNIFE

$6.95 : screwpoptool.com

Missouri inventor Brett Fischer founded Screwpop Tools, LLC, in early 2009. Their flagship product was a combination screwdriver, nut driver, and bottle opener designed to hang on a key ring. Fischer sent me one in early 2010, and it's been on my keychain ever since, weathering five straight years of pocket wear and regular use like a champ. Meanwhile, Screwpop's "keychain toolbox" line has grown to include pliers, a wrench, a stash box, a lighter holder, and a flashlight.

The utility knife — which safely corrals a pair of standard contractor blades until you need them, then locks them securely in cutting position when you do — is the latest addition to the fold. None of these tools costs more than seven bucks, and all feature Screwpop's functional, reliable, sturdy, and inexpensive approach to design (plus the signature integral bottle opener to help you celebrate when the job's done). Are there better tools in the world for some of these jobs? Sure. But even the best tool is useless if it's not on hand when you need it. — *Sean Ragan*

PARK TOOL TRIPLE SPOKE WRENCH

$10 : parktool.com

I had been trying to get my girlfriend cycling for a while, so I beamed when someone offered her a free bike. Disappointingly, the back wheel was bent so badly out of shape that it lurched on each turn. I picked up a Triple Spoke Wrench from Park Tool, and turned the wobbliest bike I ever saw into an easy ride.

Small enough to pack in your kit and cheap enough to keep a spare at home, it felt so comfortable that I don't see myself going fancier any time soon. Handling three of the typical sizes of spoke nipples means it will work for her new bike, my old bike, and probably any unforeseen bike.

It's a chunk of metal that can outlast me. — *Sam Freeman*

TOOLBOX

LOCTITE 425 ASSURE PLASTIC FASTENER THREADLOCKER

$20 per 20g (0.71 oz.) : henkelna.com

There are several ways to prevent your robot's fasteners from shaking loose. One option is to use locking washers. Another is to use nylon lock nuts. But what if you're working with machine screws that thread right into laser-cut plastic? Your options become much more limited.

Threadlocker is a good choice for when you can't or don't want to use special washers or nuts, but there's a catch — a popular formulation (Loctite 242) is not considered safe for use on plastic parts, which could crack with time. In that case, you'll need Loctite's 425 formulation, which is technically an instant adhesive. It's cyanoacrylate-based, which makes it akin to super glue.

Loctite's 425 Assure threadlocker is safe for use on plastic materials, and since it is a low-strength adhesive you can still remove treated fasteners when it's time to service your bot. According to Loctite, another popular application is to tamper-proof the heads of screws or potentiometers.

— Stuart Deutsch

TIP

This is not something to buy "just in case" — each bottle has an expiration date.

ACTOBOTICS BOGIE RUNT ROVER

$70 : servocity.com

Want to make a rolling robot that can climb obstacles? Your best bet is a rover chassis like the Bogie, a robust and agile little roller that is ready for your microcontroller and sensor package. It consists of a 6"x9" ABS plate studded with mounting holes, with six independently powered wheels providing propulsion. That's right, you get six motors with this kit — though that's it for electronics. Its namesake rocker-bogie suspension allows the robot to roll over obstructions with ease. It sports 5" of clearance beneath the chassis.

The Bogie doesn't exist in a vacuum. Actobotics robot parts (sold at SparkFun and ServoCity) can be attached natively to the chassis, or you can always drill into the ABS to add whatever you want.

— John Baichtal

MINNOWBOARD MAX

$145 : minnowboard.org

Lesser known than the Intel Galileo or Edison development boards, the MinnowBoard Max might just be the best Intel board for your next project. The Max is Intel's most open-source board and available in two hardware configurations: powerful and more powerful. The latter variant packs a 1.33 dual-core 64-bit Intel Atom processor with 2Gb of DDR3 RAM, and a true 1Gb Ethernet — a rarity in small, inexpensive development boards — into a 2.9" x 3.9" board.

The Max sets a high standard for mainlining code to the Linux kernel, and no other development board that I'm aware of is in the build configuration for Android Open Source Project. Adding to the allure are the open-source video drivers, SATA, and PCIe connectors. This is a serious board, for serious development.

— David Scheltema

PANNIERS, WHY USE THEM?

Here in the Bay Area, our staff loves to ride for many reasons. Recreation, commuting, or cost savings, we all needed the same thing: durable, weatherproof bags to protect our tools, tech, and trappings.

We picked a few of our favorites to share with you — or you can make your own!

ORETLIEB BACK-ROLLER CLASSIC PAIR

SHOWN WITH SEPARATE BACKPACK CARRYING SYSTEM

$185 for pair; $37 backpack system :
ortliebusa.com

Tried and true among bikers, these Oretlieb panniers are a staff favorite. Not only do they have one of the easiest locking mechanisms we've tested, but they're durable, waterproof, and you won't have to tuck and hide straps compared to most backpack models. I love the backpack addition (they roll up and don't take up much room), but shorter staff members had issues with fit and feeling "floppy."   — *Jason Babler*

ROUTE SEVEN

$139 : northstbags.com

When you pick it up, you can just tell — it's *nice*. One of the most stylish bags out there. Reflective panels are big and well positioned. We shined our headlamp on it out camping and it looked like it was lit from within. Quite lightweight, its beauty lies in how basic it is. While it doesn't have a backpack component, it gets bonus points for big side pockets, waterproof construction and being U.S.-made in Portland, Oregon.

HIGH ROLLER 36L BACKPACK PANNIER

$129.95 : greengurugear.com

This eco-friendly, recycled bag company has one of the biggest bags we tested — and one of the heaviest. The convertible aspect is cool, but there are extra layers of material and velcro required to secure the straps when you're using it as a pannier, and hooks when you're backpacking. Great for long hauls where you need to carry a lot of gear, but less great for daily use and getting in and out of. Well-made, with an easy-expanding fold-over closure.

CONVERTIBLE WATERPROOF PANNIER BACKPACK

$79.99 : banjobrothers.com

A simple hybrid-design sets this apart: Pannier hooks on one side; backpack straps on the other. This pannier has a large, removable waterproof bag and a much-needed zippered pocket that you can access from both sides. We're fans of the big reflective stripes, and adjustable hooks. Best combination of price and features in the group and one of our favorites.

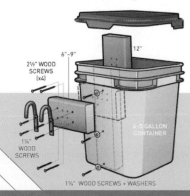

BUILD $10 BIKE PANNIERS OUT OF 5-GALLON BUCKETS

Create custom carriers from kitty litter containers with just wood, hooks, screws, washers, and a used bike tube. For step-by-step instructions, visit the project online at makezine.com/go/5gallon-bucket-panniers. Happy riding, and heavy hauling! ◐

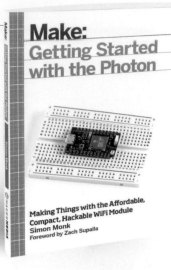

MAKE: GETTING STARTED WITH THE PHOTON
By Simon Monk
$20 : makershed.com

Priced at just $19, the Photon is an inexpensive, open source, programmable, Wi-Fi-enabled module that facilitates the building of connected projects and prototypes. Powered by an ARM Cortex-M3 microcontroller and a Broadcom Wi-Fi chip, the Photon — and its accompanying cloud platform — is designed as a ready-to-go foundation for product developers and manufacturers. Moreover, it's great for Maker projects.

The newest in our Getting Started With (GSW) series of introductory texts, *GSW Photon* teaches readers how to get started with its free development tools, deploy sketches over Wi-Fi, and build electronic projects that take advantage of the Photon's processing power, cloud platform, and input/output pins. Spark Core, the previous iteration of this platform, built by Particle (formerly Spark.io), is compatible with the Photon.

Intended for intermediate Makers who already have microcontroller experience, *GSW Photon* uses clear steps and project instructions to help Makers understand the different ways the Photon can help them explore our world. And with its low price, Photon gives plenty of reasons for Makers to experiment without having to worry about "wasting" a more expensive module.

MAKE: THE ANNOTATED BUILD-IT-YOURSELF SCIENCE LABORATORY
By Raymond E. Barrett and Windell Oskay
$25 : makershed.com

Raymond E. Barrett's *Build-It-Yourself Science Laboratory* is a classic book that took on an audacious task: To show young readers in the 1960s how to build a complete working science lab for chemistry, biology, and physics — and how to perform experiments with those tools. The experiments in this book are fearless and bold by today's standards — any number of them might never be mentioned in a modern book for young readers. Yet many from previous generations fondly remember how we as a society used to embrace scientific learning.

This new version of Barrett's book has been updated for today's world with annotations and updates from Windell Oskay of Evil Mad Scientist Laboratories, including extensive notes about modern safety practices, suggestions on where to find the parts you need, and tips for building upon Barrett's ideas with modern technology. With this book, you'll be ready to take on your own scientific explorations at school, work, or home.

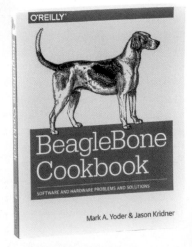

BEAGLEBONE COOKBOOK
By Mark A. Yoder and Jason Kridner
$40 : shop.oreilly.com

The *Beaglebone Cookbook* is for anyone who wants quick, simple solutions to common problems developing on any BeagleBoard platform. Both software and hardware challenges are addressed by Jason Kridner — cofounder of Beagle-Board.org — and Mark Yoder, ranging from selecting which Beagle to use in your next project to sensor and motor control solutions. More advanced topics, such as Linux kernel hacking and real-time input and output control using the BeagleBone Black's integrated programmable real-time units are explained with skill and expertise. This is a must-have volume for any fan of the BeagleBoard development platform.
—D. S.

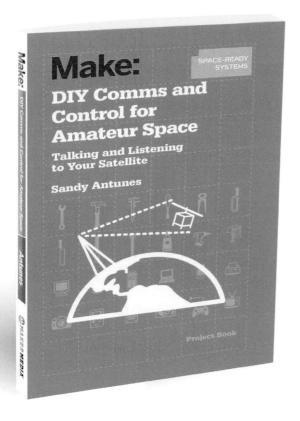

DIY COMMS AND CONTROL FOR AMATEUR SPACE: TALKING AND LISTENING TO YOUR SATELLITE
$10 : By Sandy Antunes

The fourth in professor Antunes' series for aspiring satellite builders, this book is intended as a "missing manual" to provide understanding of the entire end-to-end process necessary for accurately commanding and then receiving data from your satellite. Radio spectrum for commanding and receiving data from satellites is a shared resource with subtle hurdles. *DIY Comms and Control for Amateur Space* explores those hurdles and offers cogent, real-life advice for navigating the newest wave of amateur space exploration.

MAKE: GETTING STARTED WITH PROCESSING, 2ND EDITION
$25 : By Casey Reas and Ben Fry

Subtitled "*A Hands-On Introduction to Making Interactive Graphics,*" this text by the co-founders of the Processing Project updates their 2010 book with new material on graphics manipulation for the latest version of Processing, a simple language that lets you use code to create drawings, animation, and interactive graphics. If you're an artist looking to develop interactive graphics programs or a programmer on your way to becoming an artist and want to work with software and systems, this book is essential.

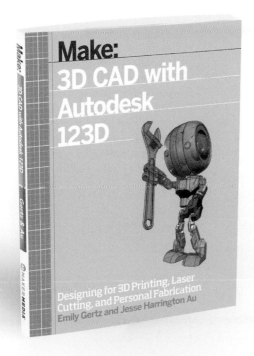

3D CAD WITH AUTODESK 123D
$30 : By Jesse Harrington Au and Emily Gertz

Autodesk's 123D suite lets you design and create real-world objects with 3D printers, laser cutters, or common materials that you assemble yourself. *3D CAD with Autodesk 123D* shows readers how to use the entire Autodesk 123D suite — including 123D Design, 123D Catch, 123D Sculpt, and 123D Make — to design, sculpt, scan, and print 3D objects with this easy-to-learn CAD interface. An ideal book for Makers, hobbyists, students, artists, and designers — including beginners — *3D CAD with Autodesk 123D* opens up the inexpensive world of personal fabrication to everyone.

PRINTRBOT METAL PLUS

The Printrbot Plus gets a metal upgrade, and the improvements are well worth the wait

WRITTEN AND PHOTOGRAPHED BY CHRIS YOHE

Printrbot Metal Plus
printrbot.com

- **Price** $1,199 single extruder $1,599 double
- **Build Volume** 254mm×254mm×254mm
- **Bed Style** Mic 6 heated aluminum bed, works with induction leveling probe
- **Temperature Control?** Yes, software adjustable on bed and extruder
- **Materials** PLA; heated bed allows for ABS. Tested with supplied Printrbot PLA, Proto-pasta carbon fiber, Taulman t-glase, SainSmart wood PLA, Taulman flexible PCTPE
- **Print Untethered?** Yes, from SD Card. OctoPrint and AstroPrint compatible
- **Onboard Controls?** External LCD add-on kit available. Otherwise control from software or OctoPrint/AstroPrint
- **Host Software** Cura and Repetier are both recommended. Cura offers the easiest starting point, and Repetier is useful for more intimate control later on
- **Slicer** Tested with Cura. Also used Slic3r with Repetier
- **OS** Windows, OSX, Linux
- **Firmware** Printrboard Firmware from Printrbot
- **Open Software?** Yes
- **Open Hardware?** Not truly open source, but CC ShareAlike 3.0

FOLLOWING THE METAL-FRAMED MAKEOVER OF ITS SMALLER SIMPLE SIBLING, the Printrbot Metal Plus transitions from a laser-cut wood frame to a new, beautiful powder-coated metal housing. The sturdy frame offers a number of upgrades over the previous generation, getting great results in our tests. It holds a z-axis that's been beefed up from the last model, and both the x- and y-axes are now running on linear rails and carts, giving the printer very smooth movement. Our unit shipped with the ceramic Ubis hot end and the aluminum extruder, as well as an aluminum build plate and heated bed that were quite responsive and quick to heat up thanks to the use of the standard ATX power supply. The pre-applied Kapton tape worked great with the flexible filament we tried, although for most of the tests we went with the commonly used blue painter's tape.

EASY SETUP, GREATER STYLE AND STABILITY

The Metal Plus ships in a robust double carton. Unboxing is simple: After removal from the inner carton and plastic wrap, one zip tie is all that remains. Keyed power connectors make setup a snap. We went from box to print in about 30 minutes, including downloading and installing Cura and following the Z-Probe tutorial. The metal frame provides a much more stable and rugged printer, helping to achieve the smooth surface finish. The powder-coated metal also puts it in line with a lot of the more design-conscious printers we've seen of late, but still keeps the design open, and not overly complicated. The included ATX power supply comes in a metal tower that matches the look of the Plus and has a built-in filament reel. The coordinating tower makes for a nice combination and helps to create a homogenous look in your work area.

FANTASTICALLY FILAMENT FRIENDLY

The included PLA printed quite well, providing a good surface finish with the default settings, even on curves. The overhang test proved to be the most challenging requiring multiple setting adjustments, and our results for the X/Y Resonance test were measured using 1mm walls.

We cranked off print after print, the vast majority of which were successful including many multi-hour torture tests. We ran a series of Makey robots in different filament materials and were surprised at how well it made the transition with just minor temperature fiddling. The aluminum extruder, with its extra support, even handled the flexible filament, although as usual, flexible material requires you to spend some time adjusting your print settings to avoid binding.

ACCESSIBLE AND UPGRADABLE

Our testing wasn't entirely problem-free, but the issues we experienced using an early machine have already been fixed in the production models along with a few other additions. Both the single and double extruder versions have heated beds, with the associated heat shielding to prevent binding due to expansion. Wiring fatigue has been reduced with a Delrin extension on the x-carriage that allows for better strain relief for the harness. We took the issues we did encounter as a chance to get inside the machine, letting us see what future maintenance and upgrades will require. The machine is laid out in a fairly straightforward manner and the wiring was easy to follow. While a machine of this size is far from simple, it is well within the wheelhouse of an average Maker, thanks to the mechanical diagrams and assembly instructions available online. This is good news for those looking to add an LCD panel, or even an extra cooling fan for the electronics, both of which already have the necessary access points built right into the base.

CONCLUSION

The Metal Plus delivers exactly what one would expect from Printrbot: a solid, robust package at a comparatively affordable price. It has a large print bed, performs well out of the box ,and can be adapted by advanced users for more exotic purposes. Really, this printer could fit into almost any user category: educators, artists, and designers, those looking to get started with a bit more capable machine, or those looking to add a larger workhorse to their stable. ✪

PRINT SCORE: 36

	1	2	3	4	5
● Accuracy				**4**	5
● Backlash					**5**
● Bridging			**3**	4	5
● Overhangs			**3**	4	5
● Fine Features				**4**	5
● Surface Curved				**4**	5
● Surface General					**5**
● Tolerance					**5**

● XY Resonance	FAIL	**PASS (2)**
● Z Resonance	FAIL	**PASS (2)**

PRO TIPS

● Make sure to check out Printrbot's great Z-Probe Calibration guides. Don't ruin your bed out of the box.
● Don't hesitate to tweak the advanced retraction settings in Cura. A little extra rise will keep your head from catching.
● If the support site falls short, head to printrbottalk.com. Also keep an eye on YouMagine for "official" DIY add-ons.
● Have fun! Whether carbon fiber, t-glase, wood, or flexible, this printer took the filaments we threw at it.

WHY TO BUY

The large build area finally meets the ease of use and setup of its smaller brother, the Metal Simple. Solid results, quick warm-up, and auto-leveling make this printer a highly recommended machine.

How'd it print?

CHRIS YOHE is a software developer by day, hardware hacker by night, and an avid 3D printing enthusiast — like many others he is slowly amassing a fleet of manufacturing minions.

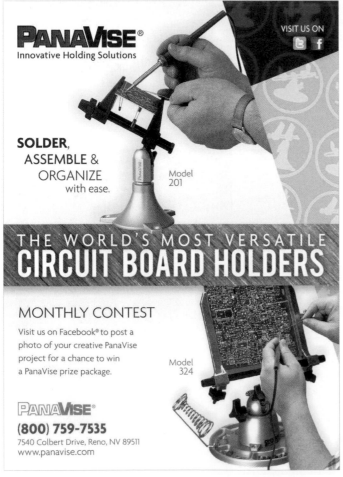

A Brief History of 21st Century BattleMechs

Written by James Burke

WELCOME TO THE INNER SPHERE SCIENTIFIC HISTORICAL ARCHIVES. TO SELECT A PREFERRED CENTURY OF INTEREST PLEASE ... you have selected 21st century: Terra.

The Republic of the Sphere's scientists have long marveled at the early innovations buried in the remains of Early Spaceflight civilizations. From the onset of classical computational devices to the splitting of the atom, our current research has made discoveries that continue to unravel the many mysteries of pre-FTL cultures. One of our finest examples currently on display could very well be the first BattleMech ever constructed. While previous archives cite the Mackie MSK-6S as the true, original BattleMech, newly discovered ComStar archives suggest that the MegaBots Mark II predates the presumed grandfather of all Mechs by more than 400 years! Tipping the scales at a featherweight of just 6 tons, the Mark II is the prototype that became the standard of 31st century combat.

This ancient marvel precedes many Mech innovations such as compact fusion reactors, energy pulse rifles, or even power armor and instead operates with fossil-fueled engines, steel armor, and compression-based weapons.

Interestingly enough, the Mark II was not an offensive war machine common to Terran battlefronts. Much like the Roman warriors of Antiquity, the Mark II was meant for arena battle, and its ballistic armaments reflected such civilized combat. Using paint-based projectiles to mark its opponents, the MegaBots faced off in amphitheaters across North America to many adoring fans during the late 21st century. Though no match for current BattleMechs, this lumbering 15-foot-tall hulk of steel represents the genesis of technologies that would evolve into the mainstay of modern warfare for over a millennia. It really makes you wonder just what other human achievements have been forever lost to time. ◙